W9-DCJ-699

VITAL
SIGNS

2006-
2007

Other Norton/Worldwatch Books

VITAL SIGNS

2006-2007

The Trends That Are Shaping Our Future

WORLDWATCH INSTITUTE

Erik Assadourian, *Project Director*

Michael Bender
Courtney Berner
Katie Carrus
Zoë Chafe
Kevin Eckerle
Christopher Flavin
Hilary French
Gary Gardner
Linda Greer

Brian Halweil
Lindsay Hower Jordan
Suzanne Hunt
Nicholas Lenssen
Zijun Li
Yingling Liu
Lisa Mastny
Danielle Nierenberg
Michael Renner

Katja Rottmann
Janet Sawin
Susan Shaheen
Hope Shand
Molly O'Meara Sheehan
Lauren Sorkin
Peter Stair
Kathy Jo Wetter
Andrew Wilkins

Linda Starke, *Editor*
Lyle Rosbotham, *Designer*

W. W. Norton & Company
New York   London

VITAL SIGNS and WORLDWATCH INSTITUTE trademarks are registered in the U.S. Patent and Trademark Office.

The views expressed are those of the authors and do not necessarily represent those of the Worldwatch Institute; of its directors, officers, or staff; or of any funders.

Composition by the Worldwatch Institute; manufacturing by Courier Westford.
Book design by Lyle Rosbotham.

ISBN-13: 978-0-393-32872-1 pbk.
ISBN-10: 0-393-32872-4 pbk.

W. W. Norton & Company, Inc.
500 Fifth Avenue, New York, NY 10110
W. W. Norton & Company Ltd.
75/76 Wells Street, London W1T 3QT

1234567890

This book is printed on recycled paper.

Worldwatch Institute Staff

Erik Assadourian
Research Associate

Courtney Berner
Development Associate
Assistant to the President

Lori A. Brown
Research Librarian

Lila Buckley
China Director

Zoë Chafe
Staff Researcher

Steve Conklin
Web Manager

Barbara Fallin
Director of Finance and
Administration

Christopher Flavin
President

Hilary French
Senior Advisor for Programs

Gary Gardner
Director of Research

Joseph Gravely
Publications Fulfillment

Brian Halweil
Senior Researcher

Alano Herro
Staff Writer

John Holman
Director of Development

Suzanne Hunt
Biofuels Program Manager

Lisa Mastny
Senior Editor

Danielle Nierenberg
Research Associate

Laura Parr
Development Assistant

Tom Prugh
Editor, World Watch

Mei Qin
Communications Officer, Beijing

Darcey Rakestraw
Communications Manager

Mary Redfern
Foundations Manager

Michael Renner
Senior Researcher

Lyle Rosbotham
Art Director

Janet Sawin
Senior Researcher

Molly O'Meara Sheehan
Senior Researcher

Patricia Shyne
Director of Publications
and Marketing

Lauren Sorkin
Programs Associate

Peter Stair
Research Assistant

Georgia Sullivan
Vice President

Andrew Wilkins
Administrative Assistant

Worldwatch Fellows

Molly Aeck
Senior Fellow

Chris Bright
Senior Fellow

Seth Dunn
Senior Fellow

David Hales
Counsel

Eric Martinot
Senior Fellow

Mia McDonald
Senior Fellow

Zijun Li
China Fellow

Yingling Liu
China Fellow

Sandra Postel
Senior Fellow

Payal Sampat
Senior Fellow

Victor Vovk
Senior Fellow

Contents

PART ONE: Key Indicators

PART TWO: Special Features

Acknowledgments

"Human activity is putting such strain on the natural functions of Earth that the ability of the planet's ecosystems to sustain future generations can no longer be taken for granted." These were the words of the Board of Directors of the Millennium Ecosystem Assessment, a four-year analysis of the world's ecosystems written by over 1,300 scientists. It is becoming ever more apparent that human society has a rapidly shrinking window of time to alter its path.

The trends documented in this short, accessible book make this reality all too apparent. Temperatures are rising, forests are shrinking, and populations of humans and their vehicles are growing, as is the need for food, energy, metal, and timber. We hope this compilation will encourage all of us as producers and consumers, as educators and students, as policymakers and citizens, and as community and business leaders to give priority to creating a sustainable world—one in which our children can enjoy as enriching lives as some of us have had.

The process of writing *Vital Signs* is a complex one, drawing on the support of thousands of Worldwatch friends. To investigate such a broad array of trends, we depend on the guidance of numerous experts who offer comments on our drafts and supply the data that provide the foundations for each trend. This year we would especially like to thank Otto Beaujon, Colin Couchman, Tom Damassa, Serena Fortuna, Pat Franklin, Jenny Gitlitz, Mark Haltmeier, Mario Hartloper, Frank Jamerson, Rachel Kaufman, David Lennett, Petra Löw, Mette Løyche Wilke, Birger Madsen, Eric Martinot, Peter Maxson, Paul Maycock, Iain McGhee, Pat Mooney, Mika Ohbayashi, Olexi Pasyuk, David Pimental, Steven Piper, Patricia Plunkert, Alfredo Quarto, Silvia Ribeiro, David Roodman, Mycle Schneider, Wolfgang Schreiber, Paul Scott, Vladimir Slivyak, Asa Tapley, Jim Thomas, Jay Townley, Wayne Wagner, Rasna Warah, Katherine Laura Watts, Kamill Wipyewski, Angelika Wirtz, and Gregor Wolbring.

We would also like to give a special acknowledgment to the contributions of Tim Whorf, a staff research associate with the Carbon Dioxide Research Group at the Scripps Institution of Oceanography, who passed away this last year. Tim provided climate data to Worldwatch over the past eight years. He and his important work will be missed.

This year we drew on the expertise of several researchers beyond the walls of the Institute. Former Worldwatch researcher Nick Lenssen continued to track the growth of nuclear energy. Recent Worldwatch interns Katie Carrus, Lindsay Hower Jordan, and Katja Rottmann contributed three articles. Linda Greer of the Natural Resources Defense Council and Michael Bender of the Mercury Policy Project provided an overview of global mercury usage. Hope Shand and Kathy Jo Wetter of the ETC Group investigated the growth of nanotechnology. Kevin Eckerle, a consultant at the U.N. Environment Programme at the time, described plant diversity trends. And Susan Shaheen, program leader at Partners for Advanced Transit and Highways at the University of California Berkeley, examined the recent expansion in global car-sharing services.

Of course, every good book needs a good publisher. We are grateful for the continued efforts of our longtime publisher, W. W. Norton & Company, and especially for the help provided by Amy Cherry, Leo Wiegman, and Anna Oler. It is their commitment that helps transform *Vital Signs* from bits and bytes to this volume, found in bookstores and classrooms across the United States.

We are also lucky enough to have a committed group of international partners who work diligently to produce *Vital Signs* outside the United States. For their considerable help in translating, publishing, and promoting recent editions, we thank Soki Oda of Worldwatch Japan, Lluis Garcia Petit and Sergi Rovira at Centro UNESCO de Catalunya in Spain, Jonathan Sinclair Wilson and Jon Raeside at Earthscan in the United Kingdom, and Eduardo Athayde in Brazil.

Now let us turn to a very important collection of people and institutions: those who allow Worldwatch to keep working. Each year a far-sighted group of foundations and governments provide us with financial support. Thanks especially to the Blue Moon Fund, the Ford Foundation, the German Government, The Goldman Environmental Prize/Richard & Rhoda Goldman Fund, the W. K. Kellogg Foundation, the Steven C. Leuthold Family Foundation, the Merck Family Fund, the Noble Venture Fund/Community Foundation Serving Boulder County, the Norwegian Royal Ministry of Foreign Affairs, The Overbrook Foundation, the David and Lucile Packard Foundation, the V. Kann Rasmussen Foundation, the Rockefeller Brothers Fund, The Shared Earth Foundation, The Shenandoah Foundation, the Taupo Community Fund of the Tides Foundation, the United Nations Population Fund, the Wallace Genetic Foundation, Inc., the Wallace Global Fund, the Johanette Wallerstein Institute, and the Winslow Foundation.

Other important contributors include the more than 3,500 Friends of Worldwatch—individuals whose commitment to the Institute has made our work possible. A special word of gratitude goes to our Council of Sponsors, who have played a central role in sustaining us for many years: Adam and Rachel Albright, Tom and Cathy Crain, and Timothy and Wren Wirth. And we thank the Worldwatch Board of Directors, an exceptional group of people whose guidance and leadership make our work better every year.

Within the walls of Worldwatch, many individuals help each year to make *Vital Signs* possible. Patricia Shyne, our Publications Director, works with our publisher and international partners to spread the book across the globe. Our development staff, consisting of Georgia Sullivan, John Holman, Mary Redfern, Laura Parr, and Courtney Berner, play a critical role in cultivating support for Worldwatch's essential work. Behind the scenes, we receive unflappable support from Director of Finance and Administration Barbara Fallin and good-humored assistance from Mail and Publication Fulfillment Coordinator Joseph Gravely.

Our communications team—Darcey Rakestraw and Drew Wilkins—works diligently to bring *Vital Signs* and other Worldwatch publications to new audiences every day. We also added a new face to the Institute just before publishing *Vital Signs*—Alana Herro, a staff writer for a new environmental news service that we recently launched. *World Watch* Magazine Editorial Director Tom Prugh and Senior Editor Lisa Mastny plan new and interesting magazine issues that inspire and inform us with new stories, while Research Librarian Lori Brown helps to gather the data that underpin our research.

At the heart of producing *Vital Signs* are two very important individuals. Linda Starke, an independent editor who has edited Worldwatch publications for more than 20 years, knows the ins and outs of this book better than anyone. Her skill and knowledge make each year's process smooth and efficient, while her wit keeps it fun. Worldwatch's Art Director Lyle Rosbotham brings the book its artistry—from the striking pictures that start the sections to the crisp color scheme and clean layout throughout.

Finally, a special acknowledgment to Steve Conklin, our Web guru, who has helped to take *Vital Signs* into the twenty-first century with the launch of *Vital Signs Online.* Now the newest trends and analyses will be housed together, ready for download, at www.worldwatch.org/vsonline. Take a look and tell us what you think at worldwatch@worldwatch.org. Thanks also to intern Sharon Kim, who was very helpful in updating and preparing the many trends for the Web version.

The future that we build can be sustainable and just. If it isn't, our children and theirs could well be reminiscing about the days when the rivers flowed freely, stomachs stayed full everyday not just on holidays, and violent tropical storms were so infrequent that they were still newsworthy. And they will ask why, why didn't anyone do anything before it was too late? So read up on the vital signs of the planet, and then get out there and protect it. Earth is the only home we've got!

Erik Assadourian
April 2006

Worldwatch Institute
1776 Massachusetts Ave., N.W.
Washington, DC 20036

Preface

Soaring commodity prices and growing signs of ecological stress are dominating the world's "vital signs" as we cross the midpoint of the first decade of the twenty-first century. The health of the global economy and the stability of nations will be shaped by our ability to address the huge imbalances in natural resource systems that now exist.

Of the 24 major ecosystem services that support the human economy—services such as providing fresh water and regulating the climate—15 are being pushed beyond their sustainable limits or are already being degraded, according to the four-year Millennium Ecosystem Assessment prepared by 1,360 scientists and released in 2005.

Some 40 percent of the world's coral reefs have been damaged or destroyed, water withdrawals from rivers and lakes have doubled since 1960, and species are becoming extinct at as much as 1,000 times the natural rate. Fish are among the world's most threatened species. The world fish catch increased nearly sevenfold between 1950 and 2000 but has now leveled off at just over 130 million tons per year. Many of the world's most important fisheries are being harvested at rates that well exceed their long-term sustainable yields. In some cases, such as Newfoundland cod, fisheries have collapsed in the face of overfishing, land-based pollution, and other stresses.

This is what scientists call a nonlinear change—one that is abrupt and potentially irreversible. Research shows that ecosystems can be overexploited for long periods of time while showing relatively little effect. But when these systems reach a "tipping point," they collapse rapidly—with far-reaching implications for all who depend on them.

Abrupt change was much in evidence in southern Louisiana and Mississippi in 2005. For decades, the flow of the Mississippi River was altered, the wetlands at its mouth were destroyed, and massive amounts of water and oil were extracted from beneath the delta. These "improvements" to nature were considered essential for economic development—helping the region to become not only a great artistic and cultural center but the hub of the U.S. petroleum industry. Few noticed that the destruction of natural systems had left New Orleans as vulnerable as a sword-wielding soldier on a high-tech battlefield. A city that was above sea level when the first settlers arrived in the eighteenth century was as much as a meter below that level when the hurricane season began in 2005.

When Hurricane Katrina came ashore on August 29th, it took less than 48 hours for much of New Orleans and coastal Louisiana and Mississippi to be destroyed. The estimated $129 billion in economic damage from this one storm exceeded annual losses from all weather-related disasters worldwide in any previous year. And Hurricane Katrina was not an isolated anomaly. The average annual losses from weather-related disasters has increased more than fourfold since the early 1980s, while the number of people affected by these catastrophes has jumped from an average of 97 million a year in the early 1980s to 260 million a year since 2001.

The mounting toll from such disasters has several causes, including rapid growth in the human population—now approaching 6.5 billion—and the even more dramatic growth in human numbers and settlements along coastlines and in other vulnerable areas. Climate change may also be contributing to the rising tide of disasters, according to scientific studies published in 2005. Three of the 10 strongest hurricanes ever recorded occurred in 2005, and the average intensity of hurricanes is increasing, according to recent research.

This is not surprising, since the main "fuel" that drives hurricanes is warm water. Gulf of Mexico temperatures were at record high levels in the summer of 2005—turning Hurricane Katrina in just over 48 hours from a low-level Category 1 hurricane to the strongest Atlantic storm ever recorded. (In September 2005, Hurricanes Wilma and Rita each broke Katrina's record as the strongest storm ever in that region.)

The average temperature of Earth's lower atmosphere also set a new record in 2005. Although the official temperature record extends only to 1880, climate scientists believe that these are the highest temperatures experienced since human civilization began 10,000 years ago. Further, they note that we are now within 1 degree Celsius of the highest temperature Earth has experienced in the last 1 million years—before the emergence of *Homo sapiens*. The impact of the warmer temperatures is seen most clearly in the rapid melting of glaciers around the world. Also, the area of the Arctic Ocean covered by sea ice in the summer has declined by 27 percent in the past 50 years, according to the latest estimates.

This is but a foreshadowing of what is to come: the concentration of carbon dioxide, the main greenhouse gas that is driving climate change, has reached its highest level in 600,000 years—and the annual rate of increase in these levels is accelerating, according to atmospheric measurements in 2005. This suggests that a "positive feedback loop" is coming into play, with ecological changes impeding the ability of natural systems to absorb carbon dioxide as well as causing some ecosystems to release carbon dioxide into the atmosphere. Global warming may in effect be fueling more global warming. We could be on the verge of a tipping point at which climate change shifts from a gradual process that can be forecast by computer models to one that is sudden, violent, and chaotic.

Scientists are beginning to shed their usual reserve in the face of ever-more alarming evidence. In early 2006 James Hansen, the lead climate researcher at NASA, and five other top climate scientists warned that "additional global warming of more than 1 degree C above the level of 2000, will constitute 'dangerous' climate change as judged from likely effects on sea level and extermination of species." If either the Greenland or the West Antarctic ice sheet were to melt, hundreds of millions of coastal residents would be displaced—a thousand times the scale of the New Orleans disaster. In the Shanghai metropolitan area alone, 40 million could lose their homes. And large sections of Florida would simply disappear.

In a notable media shift with important policy implications, U.S. news organizations declared in April 2005 that the debate on climate change was "over." A cover story in *Time* magazine and a full week of coverage on ABC's *World News Tonight* both acknowledged the overwhelming scientific consensus on climate change. As *Time* said, "By any measure, Earth is at the tipping point…. Suddenly and unexpectedly, the crisis is upon us."

The combination of mounting scientific evidence and revelations that ExxonMobil and other oil companies sought deliberately to muzzle government-funded climate scientists appears to have persuaded many reporters and editors to abandon their past portrayal of climate change as an unsettled scientific controversy. *Time* noted that many climate change skeptics are funded by the fossil fuel industry and that oil company lobbyists have assumed key positions in the White House and government agencies in recent years.

If melting ice and catastrophic storms are not enough to bring on an energy transition, the oil market is offering a helping hand. Oil prices in 2005 and early 2006 gyrated wildly, flirting several times with $70 a barrel, the highest prices in real terms in more than 20 years. The cause is simple: geologists are no longer finding enough oil to replace the 83 million barrels that are extracted each day. Although world oil production has not yet peaked, growth is clearly slowing—and is no longer keeping up with the demand for oil, which is growing at roughly 2 percent a year.

Since much of the remaining oil is in some of the world's least stable regions, events such as the kidnapping of oil workers by a Nigerian rebel group and tensions over Iran's nuclear program have helped spur recent spikes in the price of oil. In Senate testimony in early 2006, U.S. Secretary of State Condoleezza Rice said that nothing had taken her more by surprise since assuming office than the way that the politics of energy is warping international diplomacy, with oil wealth encouraging a host of nations to oppose U.S. policies in ways they would not have a few years ago.

In 2005, the reality of a new energy era began to sink in—from the $36 billion in profits recorded by ExxonMobil (the highest for any company in history) to the dangers of bankruptcy being faced by General Motors and by several fuel-guzzling airlines. In the United States, sales of large sports utility vehicles have plummeted, while those of hybrid electric cars have doubled in little more than a year. And in China, government leaders have responded to rising fuel prices by increasing the tax on large vehicles and mandating higher levels of efficiency.

None of this has yet been sufficient to bring energy markets into balance. But one tipping point can lead to another, and signs are now growing that the world is on the verge of an energy revolution. The already rapid growth of renewable energy industries has accelerated in the past year, with ethanol production increasing 19 percent, wind power capacity 24 percent, and solar cell production 45 percent.

The energy technology growth surge is propelled by scores of new government policies and by surging private investment. And it is attracting major commitments by multinational companies such as General Electric, Siemens, and Sharp, while also becoming one of the hottest fields for venture capitalists who are financing scores of small start-up firms. Even oil companies are getting into the act: BP and Shell are investing in solar energy and wind power.

These developments are impressive and are likely to provoke far-reaching changes in world energy markets within the next five years. But the change is still not fast enough to bring on the broader changes in the global economy that could stave off imminent ecological and economic crises. Government leaders and private citizens will have to mobilize in an unprecedented way if we are to have any chance of passing a healthy and secure world on to the next generation.

Christopher Flavin
President
Worldwatch Institute

VITAL SIGNS ONLINE

Over 100 global indicators are now available online at…

www.worldwatch.org/vsonline

Categories of trends:
- Food and Agriculture
- Energy and Climate
- Economy
- Transportation and Communications
- Health and Social
- Conflict and Peace
- Environment

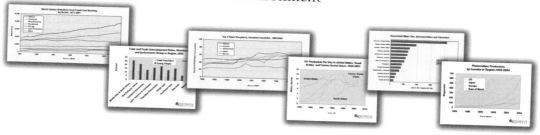

For each trend you can download the analysis and the data, charts, and graphs both in Excel spreadsheet format and in PowerPoint so you can add them to your own reports and presentations about the trends shaping our future. Get one trend or several from the seven different categories.

TECHNICAL NOTE

Units of measure throughout this book are metric unless common usage dictates otherwise. Historical population data used in per capita calculations are from the Center for International Research at the U.S. Bureau of the Census. Historical data series in *Vital Signs* are updated in each edition, incorporating any revisions by originating organizations.

Unless otherwise noted, references to regions or groupings of countries follow definitions of the Statistics Division of the U.N. Department of Economic and Social Affairs.

Data expressed in U.S. dollars have for the most part been deflated to 2005 terms. In some cases, the original data source provided the numbers in deflated terms or supplied an appropriate deflator. Where this did not happen, the U.S. implicit gross national product deflator from the U.S. Department of Commerce was used to represent price trends in real terms.

VITAL
SIGNS

2006-
2007

Part One

KEY INDICATORS

Food and Agricultural Trends

Catch of the day

▶ Grain Harvest Flat

▶ Meat Consumption and Output Up

▶ Fish Harvest Stable But Threatened

▶ Pesticide Trade Shows New Market Trends

For data and analysis on food and agricultural trends, including sugar consumption, coffee production, and fertilizer use, go to www.worldwatch.org/vsonline.

Grain Harvest Flat

Brian Halweil

The world's harvest of wheat, rice, corn, and other grains that make up the majority of our diets held nearly steady in 2005 at 2,015 million tons.[1] (See Figure 1.) The harvest dipped just 2.5 percent from the record harvest of 2004, which broke the 2-billion-ton mark for the first time in history.[2]

Grains dominate the world's diet and agricultural landscapes. They account for 47 percent of our calories and 42 percent of protein.[3] Farmers plant grains on half the world's cropland and on nearly two thirds of the irrigated land.[4] Maize (corn) follows soybeans as the dominant type of genetically modified crop, with 21.2 million hectares of corn engineered to resist herbicides or produce insecticides in 2005, up from just 300,000 hectares in 1996.[5] (Genetically modified rice has been introduced on a small scale, and scientists are working on engineered wheat.)

LINKS pp. 24, 28,104

In 2005, China, India, the United States, and the European Union—the world's major grain producers, which together account for 60 percent of global output—all had normal harvests.[6] Drought caused substantial crop losses in North Africa and South America, although this did not have a major impact on the global figure.[7]

The harvest in all of Africa increased by 6 percent over 2004's figure, buoyed by a 16-percent jump in the West Africa harvest, from 36.4 million to 46.3 million tons.[8] In contrast, North Africa—typically the continent's major grain producer—registered a 16-percent decline, from 36.3 million to 30.6 million.[9] The continent's largest producer of wheat and rice, Egypt, was insulated from these losses by its heavy use of irrigation; it actually recorded a slighty increased harvest.[10]

South America, a growing exporter of corn, also suffered from poor weather in 2005. Both Argentina and Brazil, the region's top producers, planted significantly less land in grains due to drought during planting time.[11] The region harvested almost 21 percent less grain than during the record 2004 year and about 8 percent below the region's five-year average.[12]

A favorable monsoon from India to South-

east Asia produced a record rice crop in Asia, where 90 percent of the world's rice is grown.[13] More than half the people in the world eat rice on a daily basis, making it more important for humanity than wheat or corn.[14]

The global corn harvest declined by more than 4 percent from 2004's record 724 million tons to 692 million tons (see Figure 2): a hot, dry stretch cut the harvest by 7 percent in the U.S. midwestern Corn Belt, which accounts for 42 percent of global production and 58 percent of global corn trade.[15] And in Europe, the third most important region, excess rain reduced the corn harvest by almost 15 percent.[16] Farmers in China, the world's second major producer, harvested less corn as well.[17]

Wheat also declined slightly from last year's record, as the same weather that damaged corn crops in Europe and North America hurt wheat crops.[18] In Asia, the world's largest wheat region, farmers in India and China both harvested more wheat than in 2004.[19] The 2005 harvest of 626 million tons is about 1 percent below the record in 2004.[20]

The wheat harvest increased substantially in the hungriest nations, a group of low-income, food-deficit countries.[21] Nonetheless, no immediate effect on hunger was noticeable. According to the latest estimate, 852 million people are hungry every day and nearly 6 million children die each year from diseases related to malnutrition, a figure that roughly equals the entire preschool population of Japan.[22]

Despite a 4-percent dip in the per capita production of grain, from 324 kilograms to 312 kilos (see Figure 3), per capita consumption actually increased slightly in 2005 as countries drew down stocks and as high prices meant that slightly less grain became animal feed.[23]

World cereal stocks continue their long-term decrease—they now stand at about 22 percent of annual consumption, the equivalent of about 80 days of consumption.[24] That is a drop of about 40 days just since the late 1990s.[25]

Figure 1. World Grain Production, 1961–2005

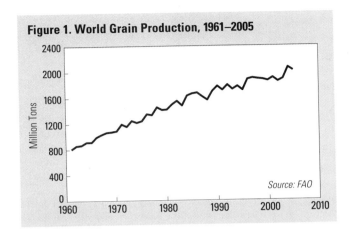

Figure 2. Corn, Wheat, and Rice Production, 1961–2005

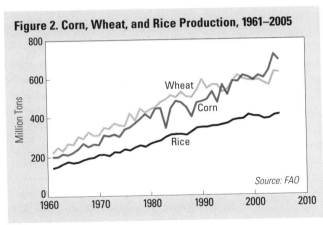

Figure 3. World Grain Production Per Person, 1961–2005

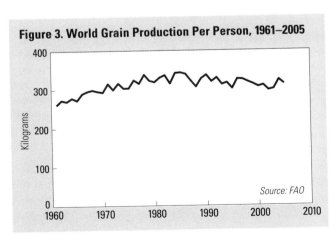

World Grain Production, 1961–2005

Year	Total	Per Person
	(million tons)	(kilograms)
1961	805	261
1965	914	273
1966	992	290
1967	1,032	296
1968	1,065	299
1969	1,073	296
1970	1,087	293
1971	1,194	315
1972	1,156	300
1973	1,246	316
1974	1,216	303
1975	1,241	303
1976	1,348	324
1977	1,333	315
1978	1,454	338
1979	1,413	323
1980	1,418	319
1981	1,496	330
1982	1,552	337
1983	1,478	315
1984	1,632	342
1985	1,665	343
1986	1,678	340
1987	1,618	322
1988	1,565	306
1989	1,700	327
1990	1,779	337
1991	1,717	320
1992	1,798	330
1993	1,727	312
1994	1,777	317
1995	1,715	301
1996	1,883	326
1997	1,903	325
1998	1,891	319
1999	1,882	313
2000	1,861	306
2001	1,909	310
2002	1,850	297
2003	1,891	300
2004	2,066	324
2005 (prel)	2,015	312

Source: FAO.

Meat Consumption and Output Up

Danielle Nierenberg

Worldwide meat production grew in 2005, with 265 million tons produced, a 2.5-percent increase from 2004.[1] (See Figure 1.) Global meat production has increased fivefold since the 1950s and more than doubled since the 1970s.[2]

Consumption of meat and other animal products also continues to rise, especially in developing countries. From the early 1970s to the mid-1990s, meat consumption in developing countries grew by 70 million tons—almost three times as much as in industrial countries.[3] Nearly 60 percent of the meat consumed worldwide is eaten by people in developing countries, up from 43 percent in the 1990s.[4] Yet it is consumers in the industrial world who still eat the most meat per person, at 85 kilograms a year.[5] (See Figure 2). By comparison, the average person in the developing world eats 31 kilograms a meat per year, although that is nearly double the level in 1990.[6]

pp. 54, 102,120

Global poultry output rose by 4 percent in 2005 to more than 81 million tons, up nearly 3 percent from the previous year.[7] (See Figure 3.) This growth occurred even as avian flu continued spreading across Asia and into Europe and Africa. As of mid-April 2006, this disease had killed nearly 110 people and led to the slaughter of millions of chickens.[8]

Beef production topped 60 million tons in 2005.[9] Traditionally, beef production has been highest in industrial countries, but developing countries now produce more than half of the world's beef.[10] China, Indonesia, and Viet Nam, for example, have all experienced gains in output, thanks to high prices for beef, while demand for beef exports have pushed up production in South America.[11]

Pork production grew in 2005 to nearly 103 million tons from 100 million tons in 2004.[12] China continues to be the world's largest producer and consumer of pork and is home to more than half of the world's 950 million pigs.[13] Demand in other countries, including Mexico, Viet Nam, and other Asian countries, pushed the developing world's share of pork production to 62 percent, up from 55 percent just a decade earlier.[14]

Factory farming, or industrial animal production, continues to grow, helping to increase production. Industrial systems today are the source of 74 percent of the world's poultry products, 50 percent of all pork, 43 percent of beef, and 68 percent of eggs.[15]

Industrial countries dominate production, but it is in the developing world where livestock producers are rapidly expanding and intensifying their production systems. China, for example, has an estimated 14,000 confined animal feeding operations (CAFOs), and about 15 percent of the pork and chicken in China comes from factory farms.[16] (The United States, in comparison, has 18,900 CAFOs.)[17] And in Brazil, U.S.-based Smithfield Foods and a Brazilian partner have built one of the world's largest pig farms, home to more than 150,000 animals.[18]

Producing this much meat, however, can entail some high costs. Factory farms are very efficient at spreading diseases, from foodborne pathogens (such as *E. coli*) to mad cow disease and avian flu. BSE (bovine spongiform encephalopathy), or mad cow disease, was first discovered in the United Kingdom in the 1990s and continues to be found in cows in North America, Japan, and Europe. And variant Creutzfeld-Jakob disease, its human form, has killed at least 150 people.[19]

Although unsanitary conditions and the close concentration and genetic uniformity of animals in factory farms may have helped lead to the emergence and spread of avian flu, officials from the Food and Agriculture Organization and the World Health Organization are recommending, at least in the short term, moving all poultry production to commercial farms and eliminating backyard production.[20] In April 2005, Viet Nam banned live poultry markets and began requiring farms to convert to factory-style farming methods in Ho Chi Minh City, Hanoi, and other cities.[21] Although these restrictions will drive thousands of small producers out of business and eliminate traditional methods of food production, it may be the only way, at least for now, for some countries to prevent the further spread of avian flu and its threat to human health.

Figure 1. World Meat Production, 1961–2005

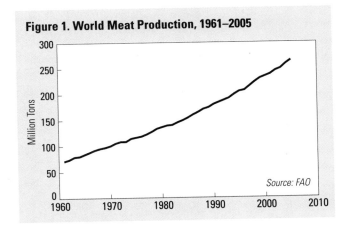

Source: FAO

Figure 2. Meat Consumption Per Person, World, Industrial Countries, and Developing Countries, 1961–2005

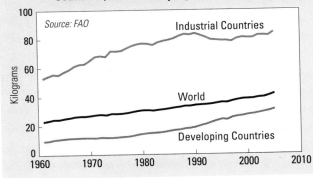

Source: FAO

Industrial Countries

World

Developing Countries

Figure 3. World Meat Production by Source, 2005

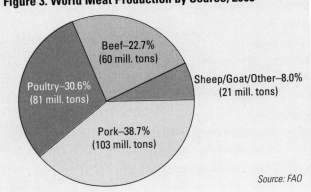

Beef–22.7%
(60 mill. tons)

Sheep/Goat/Other–8.0%
(21 mill. tons)

Poultry–30.6%
(81 mill. tons)

Pork–38.7%
(103 mill. tons)

Source: FAO

World Meat Production, 1961–2005

Year	Total	Per Person
	(million tons)	(kilograms)
1961	71	23.1
1965	84	25.2
1966	88	25.7
1967	92	26.4
1968	95	26.7
1969	97	26.7
1970	100	27.1
1971	105	27.6
1972	108	27.9
1973	108	27.5
1974	114	28.4
1975	116	28.3
1976	118	28.5
1977	122	28.9
1978	127	29.6
1979	133	30.3
1980	136	30.7
1981	139	30.7
1982	140	30.4
1983	145	30.9
1984	149	31.3
1985	154	31.8
1986	160	32.4
1987	165	32.8
1988	171	33.5
1989	174	33.5
1990	180	34.1
1991	184	34.3
1992	188	34.5
1993	192	34.8
1994	199	35.4
1995	205	36.0
1996	207	35.9
1997	215	36.9
1998	223	37.7
1999	230	38.3
2000	234	38.4
2001	238	38.6
2002	245	39.3
2003	249	39.5
2004	258	40.6
2005 (prel)	265	41.9

Source: FAO.

Fish Harvest Stable But Threatened

Brian Halweil

The world's fish farmers and fishing fleets harvested 132.5 million tons of seafood in 2003 (the last year for which data are available), just slightly less than 2002's record of 133 million tons.[1] (See Figure 1.) The amount of seafood available for each person on the planet declined slightly to 21 kilograms, down from a high of 21.5 in 2000.[2] (See Figure 2.)

The global fish supply was buoyed entirely by increased production from fish farms, since wild harvests from streams, lakes, bays, and oceans dropped from 81.4 million tons in 2002 to 77.7 million tons in 2003.[3] In fact, the gap between wild harvests and fish raised on farms continues to narrow. Since 1997, wild harvests have fallen 13 percent from the peak of roughly 87 million tons, while fish farming's harvest jumped more than 50 percent, from 35.8 million tons to 54.8 million tons.[4]

LINKS pp. 92, 106

Worldwide, fishing is highly concentrated. Of the estimated 30,000 existing fish species, only 1,000 are eaten by humans, and a small number of these make up the bulk of the catch.[5] For instance, Alaska pollack, Peruvian anchovy, Atlantic bluefin tuna, and Chilean jack mackerel account for about 13 percent of the global wild catch (about 11 million tons).[6] Fewer than 10 species—mainly carp, catfish, tilapia, and salmon—dominate global aquaculture.[7]

And just seven nations—China (47.3 million tons), Peru (6.1 million), India (5.9 million), Indonesia (5.7 million), the United States and Japan (5.5 million each), and Chile (4.2 million)—take in nearly two thirds of the global total.[8]

Fishers in developing countries catch three out of four wild fish, by weight.[9] People in the developing world also eat most of the world's fish, thanks to the larger populations there, although they eat much less per capita: 14.2 kilos per year compared with 24 kilos in the industrial world.[10] For nearly 1 billion people, mostly in Asia, fish supplies 30 percent of their protein, versus just 6 percent of protein worldwide.[11]

Promoted partly by improved refrigeration on fishing boats and rapid transportation, trade in seafood has soared in recent decades. Since 1961, the volume of trade has jumped fourfold to 26.8 million tons, while the value has jumped ninefold to $61 billion (in 2005 dollars).[12] (See Figure 3.) Worldwide, fish processors exported 10.8 million tons of frozen fish in 2001, over 22 times more than in 1961, with frozen shrimp and squid experiencing particularly rapid growth.[13]

While fisheries scientists have repeatedly confirmed that the major ocean fish stocks are overtapped, recent studies also show that freshwater fish—accounting for 8.7 million tons, excluding fish farming and sport fishing—are joining marine species among populations threatened by overfishing, pollution, and destruction of habitat.[14] Large species like the Mekong catfish, the Murray cod, and the Great Lakes sturgeon are all vulnerable to extinction.[15] Total catches are actually increasing in developing nations like India, Bangladesh, Egypt, Tanzania, and Uganda that depend heavily on freshwater species for food and jobs, even though the size and quality of fish is declining and other indicators show ecosystem stress.[16]

But conscientious chefs and seafood eaters are helping to encourage better fishing practices by shunning fish under threat of collapse. For instance, public outcry in late 2004 led to bans on serving shark fin soup at several high-profile locations, including Hong Kong's premier university and Hong Kong Disney.[17] Sharks, whose numbers have declined by more than 70 percent worldwide, are hunted primarily for their fins, which can sell for up to $700 per kilo and which constitute the central ingredient in this soup prized by Chinese populations around the world.[18]

A more ambitious strategy for rebuilding the world's fish stocks is the large-scale use of marine reserves—areas where fishing would be banned to allow fish to spawn and mature uninhibited. Evidence shows that fish populations recover rapidly in such reserves and that nearby fish catches and fish sizes increase dramatically after the reserves are set up.[19] A recent study estimated that establishing reserves for all the world's major fisheries would cost $5–19 billion each year and create about 1 million jobs.[20]

Figure 1. World Fish Harvest, 1950–2003

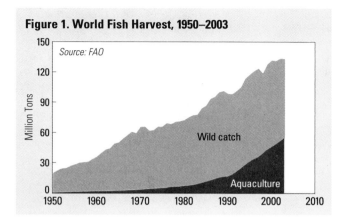

Figure 2. World Fish Harvest Per Person, 1950–2003

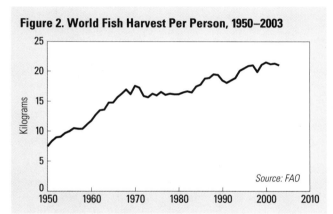

Figure 3. World Fish and Fish Products Trade, 1961–2001

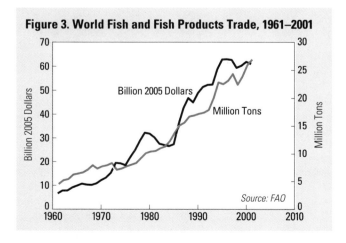

World Fish Catch and Aquaculture, 1950–2003

Year	Catch	Aquaculture
	(million tons)	
1950	19.3	0.6
1955	27.9	1.3
1960	35.5	2.0
1965	49.6	2.6
1970	65.2	3.5
1971	65.5	3.9
1972	61.4	4.3
1973	62.0	4.5
1974	65.5	4.9
1975	65.4	5.2
1976	68.9	5.4
1977	68.0	6.3
1978	70.2	6.6
1979	70.8	6.7
1980	71.9	7.3
1981	74.6	7.8
1982	76.8	8.2
1983	77.3	9.1
1984	83.6	10.2
1985	86.2	11.4
1986	92.9	12.7
1987	94.9	14.0
1988	99.5	15.5
1989	100.6	16.5
1990	97.9	16.8
1991	97.5	18.3
1992	100.6	21.2
1993	104.4	24.5
1994	112.9	27.8
1995	116.8	31.2
1996	120.4	33.8
1997	122.9	35.8
1998	118.1	39.1
1999	127.1	43.0
2000	131.0	45.7
2001	130.6	48.5
2002	133.0	51.6
2003	132.5	54.8

Source: FAO.

Pesticide Trade Shows New Market Trends Lindsay Hower Jordan

In 2004, world exports of pesticides reached $15.9 billion (in 2005 dollars), a new high in sales for this segment of the global chemical industry.[1] (See Figure 1.)

Since 1960, the pesticides trade has experienced substantial, steady growth in exports from developing countries.[2] (See Figure 2.) In 1961, these nations exported less than $1 billion of pesticides, accounting for 9 percent of world exports.[3] By 2004 the figure reached $3.3 billion, 21 percent of world exports.[4] Most notably, China has become a distinctive force in the industry, exporting $1.3 billion of pesticides—8 percent of the global total—in 2004, an increase of 155 percent since 1995, when its exports were worth $510 million.[5] (See Figure 3.)

LINKS pp. 22, 96, 98, 116

In industrial countries, although they account for 77 percent of world exports, pesticide sales have been more erratic.[6] Since 1995, for example, France's exports rose 47 percent, yielding $2.6 billion in sales in 2004, even though they dropped 13 percent between 2000 and 2001.[7] The United States recently experienced a similar dip: between 1998 and 2003, sales dropped 16.5 percent.[8] But 2004 more than compensated for this, with U.S. companies exporting $1.7 billion worth of pesticides—an 18-percent increase in a single year.[9]

France, Germany, the United States, and the United Kingdom lead the world in both exports and imports of pesticides.[10] In 2004, France exported $2.6 billion worth and brought in $1.7 billion worth; Germany exported $1.8 billion worth but imported pesticides worth $8.4 billion.[11] And the United States and the United Kingdom imported $748 million and $741 million worth in pesticides respectively.[12] China, in contrast, imported just $310 million worth of pesticides in 2004, a remarkable 45-percent decrease since 1995.[13]

Pesticide use has risen dramatically worldwide since 1961, from 0.49 kilograms per hectare to 2 kilograms in 2004.[14] The increase is attributed mainly to China's rising use of herbicides on genetically modified crops. In Germany, where the European branch of the Pesticide Action Network (PAN) is working to reduce pesticide use, application rates have remained relatively constant since 1994.[15]

According to the World Health Organization, some 3 million people a year suffer from severe pesticide poisoning.[16] The chemicals in pesticides can also contaminate drinking water when they run off farmers' fields. They are among the synthetic chemicals that are now found at unacceptable levels in the bodies of people worldwide and that can cause cancer, birth defects, and damage to the nervous system.[17] Sweden, Norway, and Denmark are implementing taxes on pesticides as a way to raise awareness about such problems and are considering legislative reforms to reduce usage.[18] Sweden has cut pesticide use by 68 percent over the past 12 years, which has helped reduced accidental poisonings by 77 percent.[19]

Alternative pest and farm management, a new approach to addressing pest problems, is being more widely adopted around the world.[20] Integrated pest management (IPM), sustainable agriculture, and organic agriculture all fall in this category of "alternatives." In a study of 22 U.S. farmers in 16 states conducted by the Natural Resources Defense Council, a comparison of alternative pest management systems (other than IPM) with conventional agricultural practices found that farmers can earn higher returns with alternative systems.[21] Similar results were found in a PAN study in the Philippines.[22]

The Rotterdam Convention on Prior Informed Consent Procedure for Certain Hazardous Chemicals and Pesticides in International Trade, which entered into force in February 2004, provides a framework by which the world can monitor and control the trade in pesticides as well as other hazardous chemicals.[23] In September 2005, delegates finalized arrangements for a secretariat and discussed how to integrate the implementation processes of this treaty, the Basel Convention on the Control of Transboundary Hazardous Waste, and the Stockholm Convention on Persistent Organic Pollutants.[24] To date, 106 countries have ratified the Rotterdam convention, including France and Germany.[25] The United States is also a signatory, but it has not yet ratified the treaty.

Figure 1. World Exports of Pesticides, 1961–2004

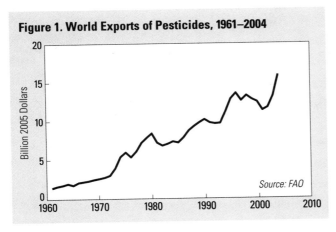

Source: FAO

Figure 2. Pesticide Trade in Industrial and Developing Countries, 1961–2004

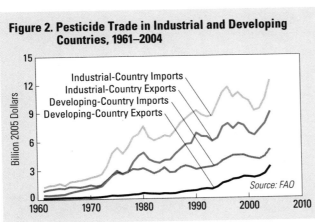

Industrial-Country Imports
Industrial-Country Exports
Developing-Country Imports
Developing-Country Exports

Source: FAO

Figure 3. Exports and Imports of Pesticides in China, 1995–2004

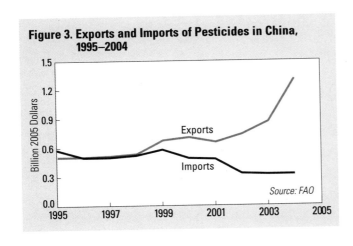

Exports

Imports

Source: FAO

World Exports of Pesticides, 1961–2004

Year	World Exports
	(billion 2005 dollars)
1961	1.4
1962	1.6
1963	1.7
1964	1.9
1965	1.7
1966	2.1
1967	2.2
1968	2.3
1969	2.4
1970	2.6
1971	2.7
1972	3.0
1973	3.9
1974	5.4
1975	6.0
1976	5.3
1977	6.1
1978	7.2
1979	7.8
1980	8.4
1981	7.2
1982	6.8
1983	7.0
1984	7.3
1985	7.2
1986	7.8
1987	8.7
1988	9.3
1989	9.8
1990	10.2
1991	9.7
1992	9.6
1993	9.7
1994	11.1
1995	12.8
1996	13.6
1997	12.6
1998	13.2
1999	12.8
2000	12.4
2001	11.3
2002	11.7
2003	13.2
2004	15.9

Source: FAO.

Energy and Climate Trends

Photodisc

Fossil fuels: high-test and regular

▶ Fossil Fuel Use Continues to Grow

▶ Nuclear Power Inches Up

▶ Wind Power Blowing Strong

▶ Solar Industry Stays Hot

▶ Biofuels Hit a Gusher

▶ Climate Change Impacts Rise

▶ Weather-related Disasters Affect Millions

▶ Hydropower Rebounds Slightly

▶ Energy Productivity Gains Slow

For data and analysis on energy and climate trends, go to www.worldwatch.org/vsonline.

Fossil Fuel Use Continues to Grow

Katja Rottmann

World oil use increased by 1.3 percent in 2005, a significant slowdown after a record-breaking rise of 3.4 percent in 2004.[1] The International Energy Agency estimates that oil demand reached 3.8 billion tons in 2005, or 83.3 million barrels a day. (See Figure 1.)[2]

The United States remained the world's largest consumer of oil, using 20.8 million barrels a day—nearly one fourth of the world total.[3] The other major oil users were Europe (15.6 million barrels daily), China (6.6 million barrels), and Japan (5.4 million barrels).[4] On a per capita basis, the United States uses two thirds more oil than Japan does and 13 times as much as China.[5]

Rising demand, combined with limited spare production capacity, pushed the average price of oil to $56 per barrel, up from $32 in 2003.[6] (See Figure 2.) This is the highest price in real terms since 1985.[7] The soaring price pushed governments to search for alternatives to oil. Thailand decided to promote domestic biofuel production, as that country imports 90 percent of its oil.[8] China passed a law to foster renewable energy.[9]

LINKS pp. 42, 44, 48, 56, 58, 64, 68

The only substantial increases in oil production in 2005 were in Russia and the Middle East. For the second year in a row, the world's spare production capacity was at near-record lows of less than 2 million barrels per day.[10]

Production in several major suppliers, including the United States, the United Kingdom, and Indonesia, is already declining.[11] Output in Russia, one of the few countries with recent significant increases, went from 6 million barrels per day in 1996 to 9.5 million barrels in 2005.[12] But this additional oil came from old fields, and during the past two years the growth in Russian oil production has begun to slow.[13]

Coal and natural gas data are not yet available for 2005, but in 2004 the use of both fuels rose.[14] (See Figure 3.) Coal use jumped 6.3 percent, reaching 2.8 billion tons of oil equivalent, largely due to a continued surge in coal-fired power plants in China, which now uses more coal than the United States, India, and Russia combined.[15] China and India together now use 42 percent of the world's coal—the fossil fuel

with the greatest impact on human health and the climate.[16] Natural gas consumption rose 3.3 percent in 2004, reaching 2.4 billion tons of oil equivalent.[17]

Nearly 80 percent of the world's energy comes from oil, coal, or gas.[18] This heavy reliance on fossil fuels puts energy security at risk. This was demonstrated by the dispute between Russia and Ukraine in December 2005. Russia stopped providing gas to its neighbor for several days due to a disagreement on prices.[19] As a result, the flow of gas to Western Europe was also briefly disrupted.[20]

The fact that much of the world's oil is produced in politically unstable regions in the Middle East, Central Asia, and Africa may put future supplies at risk. Political unrest in Nigeria in early 2006 led to attacks on pipelines and pumping stations and caused a drop of 220,000 barrels a day in output—nearly 10 percent of that country's daily exports.[21]

Natural disasters are also a risk to the oil supply infrastructure. In 2005, hurricanes Katrina and Rita shut down oil and natural gas production in the Gulf of Mexico. For a short period, about one third of U.S. refining capacity was closed.[22]

Investment bank Goldman Sachs expects oil prices to remain high during the next five to six years.[23] Higher prices could facilitate the transition to a renewable energy economy. They could also, however, present a serious risk to the global economy, hampering economic growth, particularly in developing countries.[24] Of the world's 48 poorest countries, 33 are net importers of oil; more of the half of these are completely dependant on imports.[25] Rising oil prices are draining foreign exchange reserves and adding to government debts, complicating economic development plans.

The heavy reliance on fossil fuels—identified by President George W. Bush in January 2006 as an addiction—is having other impacts as well. In the first half of 2005, for example, 2,600 workers were killed in Chinese coal mine accidents.[26] Fossil fuels are also the main force behind the increasingly urgent problem of climate change.

Figure 1. World Oil Consumption, 1950–2005

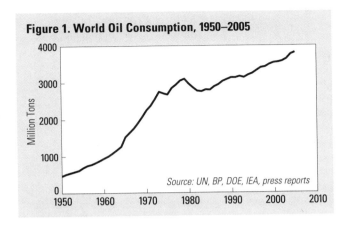

Source: UN, BP, DOE, IEA, press reports

Figure 2. Spot Crude Oil Prices 1985–2005

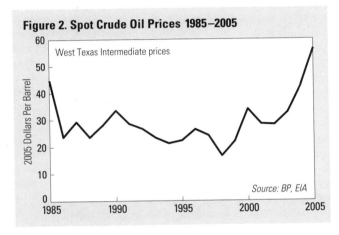

West Texas Intermediate prices

Source: BP, EIA

Figure 3. World Consumption of Coal and Natural Gas, 1950–2004

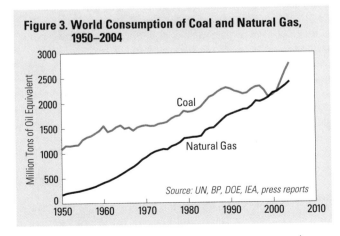

Coal

Natural Gas

Source: UN, BP, DOE, IEA, press reports

World Fossil Fuel Consumption, 1950–2005

Year	Oil	Coal	Natural Gas
	(million tons of oil equivalent)		
1950	470	1,074	171
1955	694	1,270	266
1960	951	1,544	416
1965	1,530	1,486	632
1970	2,254	1,553	924
1975	2,678	1,613	1,075
1976	2,852	1,681	1,138
1977	2,944	1,726	1,169
1978	3,055	1,744	1,216
1979	3,103	1,834	1,295
1980	2,972	1,814	1,304
1981	2,868	1,826	1,318
1982	2,776	1,863	1,322
1983	2,761	1,916	1,340
1984	2,809	2,011	1,451
1985	2,801	2,107	1,493
1986	2,893	2,143	1,504
1987	2,949	2,211	1,583
1988	3,039	2,261	1,663
1989	3,088	2,293	1,738
1990	3,136	2,270	1,774
1991	3,134	2,225	1,806
1992	3,170	2,203	1,836
1993	3,139	2,168	1,869
1994	3,204	2,186	1,877
1995	3,251	2,259	1,938
1996	3,329	2,306	2,034
1997	3,406	2,319	2,025
1998	3,426	2,239	2,060
1999	3,494	2,109	2,107
2000	3,539	2,148	2,195
2001	3,552	2,217	2,219
2002	3,581	2,413	2,282
2003	3,642	2,614	2,343
2004	3,767	2,778	2,420
2005 (prel)	3,816	n.a.	n.a.

Source: UN, BP, DOE, IEA, press reports.

Nuclear Power Inches Up

Nicholas Lenssen

Between 2004 and 2005, total installed nuclear generation capacity increased by slightly less than 1 percent, from 366,000 megawatts to more than 369,000 megawatts.[1] (See Figure 1.) The increase in 2005 came as four new reactors and one previously mothballed one were connected to the grid.[2]

Despite achieving an all-time high in terms of capacity, nuclear power's future is very uncertain. The International Energy Agency predicts that "nuclear production [will] peak around 2015 and then decline gradually."[3] Indeed, one study estimates that 80 new nuclear power plants must be ordered and built within the next 10 years in order to keep the number of operating plants constant.[4]

The pessimism for nuclear power's future partly comes from the fact that until recently only 21 reactors (with a combined capacity of 14,120 megawatts) were under active construction.[5] And in 2005, work started on just two new reactors, in Finland and Pakistan.[6] (See Figure 2.)

Two reactors were permanently closed in 2005, in Germany and Sweden, with a combined capacity of 940 megawatts.[7] Overall, some 116 reactors totaling 34,603 megawatts have been permanently closed after average lifespans of less than 21 years.[8] (See Figure 3.)

Despite the lull, new reactors are likely to be ordered over the next five years, even in the United States. Two industry consortiums there are expected to take advantage of new subsidies, including a tax credit worth 1.8¢ per kilowatt-hour for the first 6,000 megawatts of new plants and federal risk insurance of $2 billion to cover regulatory delays.[9] Beyond an expected five or so new ones, however, it is unclear whether new U.S. reactors will survive the economic test.

Canada restarted a mothballed reactor that had been shut down since 1997 and commenced work to bring another two back online.[10] But the government utility also decided not to attempt to restart two other reactors.[11]

Finland started work on the first new reactor in Europe in nearly 15 years, though the project was already behind schedule by the end of 2005.[12] France's plan to start a new reactor by 2007 is in jeopardy: a government-mandated public debate on the state utility's plans was cancelled after independent experts and environmentalists withdrew from the talks.[13] The United Kingdom expects to complete a debate this year on whether to order new reactors.[14]

Germany and Sweden continued to phase out nuclear power, and German phaseout plans survived a new government that excludes the Green Party.[15] In Spain, the parliament approved a roundtable to discuss phasing out nuclear power; the country's oldest plant is scheduled to be shut in 2006.[16] Russia has one reactor under active construction, but another five are delayed due to lack of financing.[17]

Japan has one reactor under active construction, and though plans exist on paper to add more, economics and opposition suggest that the country will continue to fall short on meeting expansion plans.[18] China has the world's most ambitious construction plans, expecting to add 31 reactors by 2020 to 9 operating ones and 2 under construction.[19] This will mean opening two new large reactors each year.

India also has ambitious plans, hoping to expand its nuclear capacity nearly 10-fold by 2022, from 3,000 megawatts operating to 29,000 megawatts.[20] But currently India only has 3,600 megawatts under construction.

Two new reactors in North Korea remain on hold, following that country's refusal to abandon its nuclear weapons program.[21] And Iran's decision to restart uranium enrichment put its plans to build new nuclear power reactors and to finish one nearing completion in jeopardy, as Iran still needs outside expertise to build reactors. In February 2006, the International Atomic Energy Agency referred Iran's nuclear program to the U.N. Security Council, which could lead to an embargo of nuclear technology and other economic sanctions against that nation.[22]

Indeed, the difficulty of limiting the proliferation of nuclear weapons remains a challenge for the civilian nuclear industry. With the underlying technical infrastructure able to support both weapons and electrons, there is no clear way to ensure nuclear energy can be developed without also building capabilities for weapons.

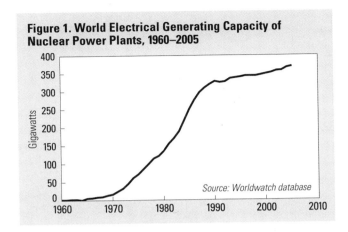

Figure 1. World Electrical Generating Capacity of Nuclear Power Plants, 1960–2005

Source: Worldwatch database

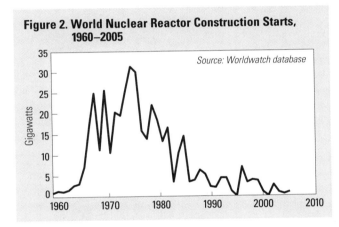

Figure 2. World Nuclear Reactor Construction Starts, 1960–2005

Source: Worldwatch database

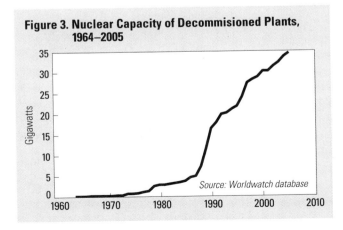

Figure 3. Nuclear Capacity of Decommisioned Plants, 1964–2005

Source: Worldwatch database

World Net Installed Electrical Generating Capacity of Nuclear Power Plants, 1960–2005

Year	Capacity
	(gigawatts)
1960	1
1965	5
1970	16
1971	24
1972	32
1973	45
1974	61
1975	71
1976	85
1977	99
1978	114
1979	121
1980	135
1981	155
1982	170
1983	189
1984	219
1985	250
1986	276
1987	297
1988	310
1989	320
1990	328
1991	325
1992	327
1993	336
1994	338
1995	340
1996	343
1997	343
1998	343
1999	346
2000	349
2001	352
2002	357
2003	358
2004	366
2005	369

Source: Worldwatch Institute database, IAEA, and press reports.

Wind Power Blowing Strong

Janet L. Sawin

Global wind power capacity jumped 24 percent in 2005, reaching nearly 60,000 megawatts at the end of the year.[1] (See Figure 1.) Wind energy generation has more than tripled since 2000, making it the world's second fastest growing energy source after solar power.[2] The estimated 11,770 megawatts of wind capacity added in 2005 was 41 percent above the previous record annual addition set in 2003.[3] (See Figure 2.)

For the first time since the early 1990s, the United States installed more wind power capacity than any other single country. An estimated 2,431 megawatts were added, bringing the U.S. total to 9,149 megawatts—trailing only Germany and Spain.[4] (See Figure 3.) The surge was driven by state policies and by congressional extension of the federal tax credit through 2007.[5] The wind turbines in place at the end of the year can meet the electricity needs of an estimated 2.3 million U.S. households.[6]

The European Union (EU) added 6,183 megawatts in 2005, bringing its total to 40,504 megawatts, exceeding the EU target of 40,000 megawatts set five years early.[7] Current European wind power capacity can produce enough electricity to meet 2.8 percent of EU demand in an average wind year.[8]

Despite a slowdown that began in 2003, Germany added 1,808 megawatts of wind power capacity for a year-end, world-leading total of 18,428 megawatts.[9] Wind power meets more than 30 percent of the annual electricity needs in three German states and about 6.6 percent for Germany as a whole.[10]

Spain added an estimated 1,764 megawatts of wind power capacity in 2005, bringing its national total to 10,027 megawatts—the world's second highest wind power capacity.[11] Spain's wind market has grown so rapidly that national targets have been updated twice since 1999, when the government first aimed for 8,900 megawatts of wind capacity by 2010.[12] The latest target calls for 20,000 megawatts by 2011, enough to generate 15 percent of Spain's electricity—up from 6.5 percent in mid-2005.[13]

Europe's other top wind markets in 2005 were Portugal, Italy, and the United Kingdom. Portugal added 500 megawatts, nearly doubling its total capacity to 1,022 megawatts.[14] Italy added 452 megawatts, lifting it to sixth place worldwide (after Denmark) with a total of 1,717 megawatts.[15] The slumbering U.K. wind market showed new signs of life, adding a record 446 megawatts to reach 1,353 megawatts at the end of 2005.[16]

Asia's wind market also saw rapid growth in 2005, representing nearly 17 percent of new capacity.[17] India added approximately 1,430 megawatts and now ranks fourth worldwide, with total wind capacity reaching an estimated 4,430 megawatts.[18] China added nearly 500 megawatts, for a total approaching 1,270 megawatts.[19] With the new Renewable Energy Law going in force since January 2006, Chinese and international companies are poised to increase wind turbine output.[20] But many investors await the reversal of a last-minute decision to promote wind energy through competitive bidding rather than feed-in tariffs.

The record growth in the United States helped create a shortage of wind turbines in 2005, preventing even greater expansion there. Past boom-and-bust cycles in the U.S. market, caused by inconsistent federal policy, have discouraged investment in domestic manufacturing of turbine components. As a result, the dramatic jump in demand in 2005 temporarily pushed up turbine prices.[21] But expectations for even stronger demand in 2006 and beyond, driven in part by soaring natural gas prices, are now driving further investment in wind turbine and component manufacturing.[22]

BTM Consult expects global wind power generating capacity to exceed 271,500 megawatts by 2015 if oil and gas prices remain high.[23] Europe's market will likely grow relatively slowly until the offshore market takes off late in this decade.[24] In the meantime, the fastest growth will likely be in North America and Asia. A 2005 study projects that U.S. capacity will exceed 28,000 megawatts by 2010.[25] Beyond that date, China may be the country with the largest potential.[26] It has the wind resources and manufacturing skills to become the world leader, but only if its renewable energy policies prove effective.

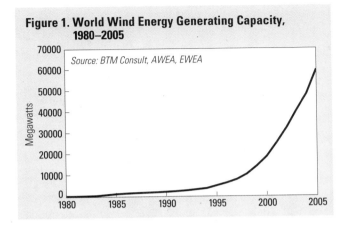

Figure 1. World Wind Energy Generating Capacity, 1980–2005

Source: BTM Consult, AWEA, EWEA

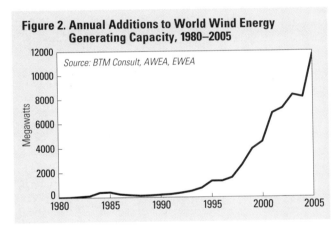

Figure 2. Annual Additions to World Wind Energy Generating Capacity, 1980–2005

Source: BTM Consult, AWEA, EWEA

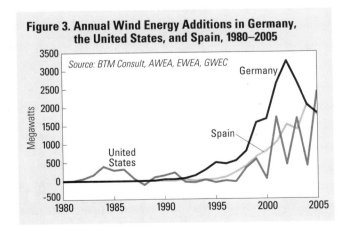

Figure 3. Annual Wind Energy Additions in Germany, the United States, and Spain, 1980–2005

Source: BTM Consult, AWEA, EWEA, GWEC

World Wind Energy Generating Capacity, Total and Annual Additions, 1980–2005

Year	Total	Annual Additions
	(megawatts)	
1980	10	5
1981	25	15
1982	90	65
1983	210	120
1984	600	390
1985	1,020	420
1986	1,270	250
1987	1,450	180
1988	1,580	130
1989	1,730	150
1990	1,930	200
1991	2,170	240
1992	2,510	340
1993	2,990	480
1994	3,490	730
1995	4,780	1,290
1996	6,070	1,290
1997	7,640	1,570
1998	10,150	2,600
1999	13,930	3,920
2000	18,450	4,500
2001	24,930	6,820
2002	32,040	7,230
2003	40,300	8,340
2004	47,910	8,150
2005 (prel)	59,600	11,770

Source: BTM Consult, AWEA, EWEA.

Solar Industry Stays Hot

Janet L. Sawin

In 2005, global production of photovoltaic (PV) cells—which generate electricity directly from sunlight—increased 45 percent to nearly 1,730 megawatts, six times the level in 2000.[1] (See Figure 1.) Cumulative production, at just over 6,090 megawatts by the end of 2005, has increased on average 33 percent a year since 2000, making solar power the world's fastest growing energy source.[2] (See Figure 2.)

Japan remains the leading PV producer.[3] The 833 megawatts manufactured there in 2005 represent an increase of 38 percent over 2004 levels and more than the entire world produced just two years earlier.[4]

LINKS pp. 36, 92

Japan's domestic PV market grew 14 percent, to 292 megawatts, with the rest of the PVs exported to a booming global market.[5] Despite a slowdown in market growth, Japan is still expected to meet a national target of 4,800 megawatts installed by 2010.[6]

Germany dominated the global market in 2005, installing an estimated 837 megawatts of new capacity.[7] High PV purchase prices have been a powerful driver of German demand, stimulating a vibrant domestic industry. The German company Q cells, established in 2003, became the world's second largest producer by 2005.[8] In early 2006, Germany's Solar World purchased the largest U.S. factory from Shell Solar.[9] Overall, Europe manufactured 452 megawatts of solar cells in 2005—up 44 percent over 2004.[10]

U.S. production rose 10 percent in 2005, to 153 megawatts.[11] The U.S. share of total production fell from 44 percent in 1996 to below 9 percent in 2005.[12] But this trend could reverse due to new state policies. In early 2006, California passed a $3.2-billion, 11-year plan to install 3,000 megawatts of solar energy.[13] As a result, manufacturers are considering major new investments there, and California could soon become a top producer of solar cells.[14]

Although all other countries manufactured less than one fifth of the world's solar cells, their PV production more than doubled to 289 megawatts in 2005.[15] China accounted for almost 42 percent of this total, with Suntech nearly tripling its capacity in 2005 alone.[16]

Demand for solar cells is booming as China attempts to meet skyrocketing energy needs. In late 2005, Shanghai launched an initiative to install PV systems on 100,000 of the city's 6 million rooftops.[17]

This boom market has created a shortage of solar-grade silicon, the primary raw material for more than 90 percent of the PV cells produced in 2005.[18] Silicon costs doubled between 2003 and 2005 due to supply constraints, causing an increase in module prices.[19] Shortages are likely to continue in 2006, limiting market growth and delaying the substantial decline in prices that scaled-up production will eventually deliver.[20] Expanded solar silicon production is expected to eliminate this bottleneck by 2007 or 2008.[21]

The market for solar heating is also expanding rapidly, with solar collectors now providing hot water to about 40 million households worldwide.[22] Total installations rose to 164 million square meters in 2005, an increase of more than 16 percent over 2004.[23] About 75 percent of these systems heat water and space (see Figure 3); the remainder are used for swimming pools.[24]

China has developed a vibrant domestic industry and had nearly 59 percent of the global heating capacity for non-pool systems in 2004.[25] The European Union ranked second in 2004, with 13 percent of the world's capacity, followed by Turkey (9 percent), Japan (7 percent), Israel (4.5 percent), Brazil (2.2 percent), and the United States (1.8 percent).[26] Israel ranks first in installations per person, with 740 square meters per 1,000 inhabitants.[27]

Germany, Europe's leading market, installed 100,000 solar thermal systems in 2005, more than 25 percent above additions in 2004.[28] It is estimated that at least 2 million Germans now live in homes with solar water and space heating, and the nation's industry employs about 12,500 people.[29] Spain could soon follow Germany's lead: in March 2006, the government approved a new building code that includes a solar heating obligation for all new and renovated buildings nationwide.[30]

Figure 1. World Annual Photovoltaic Production, 1980–2005

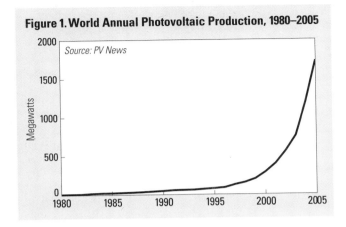

Source: PV News

Figure 2. World Cumulative Photovoltaic Production, 1980–2005

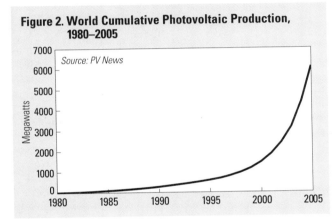

Source: PV News

Figure 3. World Cumulative Solar Heating Installations, 1995–2005

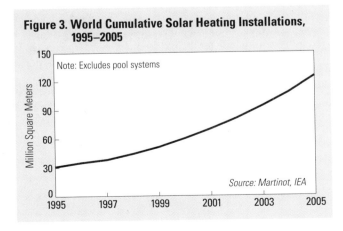

Note: Excludes pool systems

Source: Martinot, IEA

World Photovoltaic Production, 1980–2005

Year	Annual Production	Cumulative Production
	(megawatts)	
1980	6	19
1985	23	98
1990	46	273
1991	55	329
1992	58	386
1993	60	447
1994	69	516
1995	78	594
1996	89	682
1997	126	808
1998	155	963
1999	201	1,164
2000	288	1,452
2001	399	1,851
2002	560	2,411
2003	759	3,170
2004	1,195	4,365
2005 (prel)	1,727	6,092

Source: PV News.

World Solar Heating Installations, 1995–2005

Year	Annual Installations	Cumulative Installations
	(million square meters)	
1995	4	31
1996	4	35
1997	5	38
1998	6	44
1999	7	51
2000	9	60
2001	11	70
2002	12	81
2003	15	94
2004	17	108
2005 (prel)	18	125

Note: Excludes pool systems.
Source: Martinot, IEA.

Biofuels Hit a Gusher

Suzanne Hunt and Peter Stair

Production of fuel ethanol, the world's leading biofuel, jumped 19 percent in 2005 to 36.5 billion liters, continuing a growth surge that began in 2000.[1] (See Figure 1.) Ethanol, produced mainly from sugarcane and corn, accounts for more than 90 percent of the world's total biofuel production.[2] Biodiesel, derived from plant oils, is the main other type of biofuel; its output shot up 60 percent in 2005.[3] (See Figure 2.) Overall, biofuels now account for about 1 percent of the global liquid fuels market.[4]

The United States produced 16.2 billion liters of fuel ethanol in 2005, nearly surpassing Brazil to become the world's largest fuel ethanol producer.[5] These two have dominated the ethanol market since the 1980s and accounted for almost 90 percent of the output in 2005.[6]

Europe, on the other hand, had nearly 90 percent of the world biodiesel market in 2005.[7] Germany led there, producing about half the total volume, but capacity is growing rapidly in Spain, Italy, Poland, and the United Kingdom.[8] Although the United States remains a smaller producer, its biodiesel output tripled in 2005.[9]

LINKS pp. 42, 44, 64

The recent surge in biofuel production has been spurred by several developments. In the United States, ethanol is being blended with gasoline so that MTBE, a toxic additive, can be phased out.[10] In Europe, Canada, and Japan, biofuels are seen as a key way to comply with the Kyoto Protocol.[11] And throughout the world, farmers and agricultural processors have sought new markets in the face of low agricultural commodity prices and growing international pressure to reduce farm subsidies.[12]

Soaring crude oil prices have also played a big role. With oil prices over $60 per barrel for much of 2005, fuel ethanol became less expensive to produce than gasoline in many countries—even without the subsidies that have so far supported most biofuel programs.[13] In Brazil, the recent popularity of flex-fuel vehicles, which can run on a range of ethanol-gasoline blends, has let consumers take advantage of this price difference by purchasing more concentrated ethanol blends.[14] And around the world, high profit margins are driving unprecedented investment in biofuel industries—attracting newcomers ranging from British entrepeneurs to Silicon Valley venture capitalists.[15]

During 2005, a number of countries expanded or proposed new biofuel programs, including Canada, China, Colombia, the Dominican Republic, India, Malawi, the Philippines, and Thailand.[16] Such programs are particularly attractive in tropical countries with the ability to produce low-cost sugarcane ethanol and tropical oils such as palm oil.[17]

Despite regional differences in production costs, only about 10 percent of ethanol is traded internationally.[18] (By comparison, 50 percent of the world's crude oil crosses national borders.)[19] This situation is due largely to tariffs and import quotas designed to benefit domestic agricultural producers in industrial countries, where demand for biofuels is greatest.[20]

The market for biofuels has already begun to affect food markets. In Brazil, the world leader in sugar production, roughly half of the sugarcane crop goes to producing ethanol.[21] The recent rise in demand for ethanol has cut Brazil's ability to export sugar, contributing to a doubling in the world price between 2004 and 2006.[22] In Europe, 40–60 percent of the rapeseed crop is being tapped to produce biodiesel, and a shortage in oil-crushing equipment has pushed the price of rapeseed oil up.[23] About 15 percent of the U.S. corn crop was used to produce ethanol in 2005, helping farmers cope with a second consecutive bumper crop.[24]

Growing demand is also driving the development of new technologies to produce biofuels more efficiently and from a far wider array of feedstocks. Cellulosic biofuels can be produced from stalks, leaves, and other less valuable biomass fibers, and many predict they will begin to be economically viable within a decade.[25]

Biofuel production is likely to continue to skyrocket. A national Renewable Fuel Standard will nearly double ethanol production in the United States by 2012.[26] New programs elsewhere and high oil prices are likely to push production even higher.

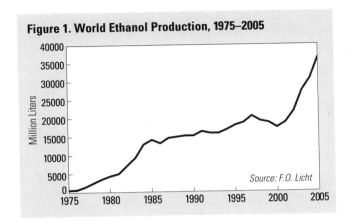

Figure 1. World Ethanol Production, 1975–2005

Source: F.O. Licht

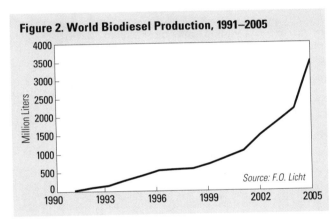

Figure 2. World Biodiesel Production, 1991–2005

Source: F.O. Licht

World Ethanol and Biodiesel Production, 1975–2005

Year	Ethanol	Biodiesel
	(million liters)	
1975	556	
1976	664	
1977	1,470	
1988	2,529	
1999	3,533	
1980	4,368	
1981	4,977	
1982	7,149	
1983	9,280	
1984	12,880	
1985	14,129	
1986	13,193	
1987	14,599	
1988	14,902	
1999	15,191	
1990	15,190	
1991	16,348	11
1992	15,850	88
1993	15,839	143
1994	16,802	283
1995	17,970	408
1996	18,688	546
1997	20,452	570
1998	19,147	587
1999	18,671	719
2000	17,315	893
2001	18,676	1,068
2002	21,715	1,488
2003	27,331	1,832
2004	30,632	2,196
2005 (prel)	36,500	3,524

Source: F.O. Licht.

Climate Change Impacts Rise

Lauren Sorkin

The average global temperature in 2005 was 14.6 degrees Celsius, making it the warmest year ever recorded on Earth's surface, according to data from NASA's Goddard Institute of Space Studies.[1] (See Figure 1.) The five warmest years since recordkeeping began in 1880 have all occurred since 1998.[2] The average global temperature has risen nearly 1 degree Celsius in the past century.[3] More than half of that warming—a rise of 0.6 degrees—has occurred in the past 30 years, meaning that this warming trend is accelerating.[4]

LINKS pp. 32, 44, 118

In 2005, the average atmospheric carbon dioxide (CO_2) concentration reached 379.6 parts per million by volume, an increase of 2.6 parts per million—0.6 percent—over the record high in 2004.[5] (See Figure 2.) Average CO_2 concentrations have climbed 20 percent since measurements began in 1959. The rise in 2005 represents the largest annual increase ever recorded.[6]

Although seemingly small, the rising temperature threatens to have profound consequences in the years ahead. Already, in September 2005 sea ice in the northern hemisphere was at its lowest levels in recorded history.[7] Greenland's glaciers lost nearly 53 cubic miles of ice in 2005 alone.[8] During the summer, heat waves kept temperatures above 38 degrees Celsius (100 degrees Fahrenheit) for 39 consecutive days in Arizona, and much of Europe was hit with forest fires, followed by torrential rains and severe flooding.[9] Abnormal algal blooms cost U.S. Gulf Coast residents $3 million a week in lost revenues from tourism, fisheries, restaurants, and related activities.[10] Likewise, Kuala Lampur experienced a downturn in the tourism industry and was forced to shut its largest harbor as a result of wild fires brought on by drought and extreme heat.[11]

Already, climate change is forcing entire communities to move or risk losing their livelihoods. For example, changing weather patterns are believed to be responsible for decreasing rainfall in the Gobi Desert that has helped it expand by 26,000 square kilometers a year and forced tens of millions of Chinese farmers to retreat.[12] Inuit natives in communities from Canada, Greenland, Alaska, and northern Russia were forced to move northward to follow prey in 2005 as a result of the warmest winter on record in the Arctic region.[13]

There is broad consensus in the scientific community that the rise in global temperatures is due to emissions of greenhouse gases such as CO_2—the primary source of which is the burning of fossil fuels.[14] CO_2 emissions from fossil fuels increased a record 4.5 percent in 2004, to 7.57 billion tons. (See Figure 3.) Research that measures gases present in ancient Antarctic ice cores has revealed that CO_2 levels are now 27 percent higher than at any point up until the start of the Industrial Revolution.[15]

The United States, with about 5 percent of the world's population, accounts for the largest share of CO_2 emissions from human activities—25 percent of global emissions.[16] China is the second largest emitter, although emissions per person there are far lower than in the United States.[17] The largest sources of U.S. emissions are coal-burning power plants, followed by automobiles.[18] Globally, transport-sector emissions are growing most rapidly, due to dramatic rises in car ownership in developing countries.[19]

Some experts fear that climate change is already set in motion and will be difficult to reverse and that this will lead to more severe storms, drastic reductions in agricultural yields, biodiversity loss, and threats to human health—the effects of which tend to hit lower-income communities more than richer ones.[20]

In 2005, efforts to address climate change globally met with varying degrees of success. To meet its commitment under the Kyoto Protocol to cut greenhouse gas emissions by 8 percent, Europe launched an Emissions Trading Scheme in January 2005.[21] Although this does not yet include several carbon-intensive industries or the transport sector, total CO_2 trade in the inaugural year was worth 9.4 billion euros, and the volume of CO_2 traded reached 799 million tons.[22] While the United States remains outside the Kyoto Protocol, a growing number of state and city governments are taking action to reduce emissions.[23]

Figure 1 . Global Average Land-Ocean Temperature at Earth's Surface, 1880–2005

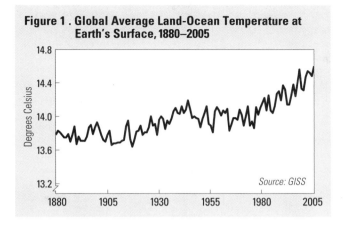

Source: GISS

Figure 2. Atmospheric Concentrations of Carbon Dioxide, 1960–2005

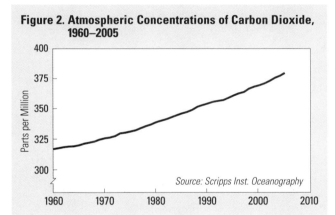

Source: Scripps Inst. Oceanography

Figure 3. Carbon Emissions from Fossil Fuel Burning, 1950–2004

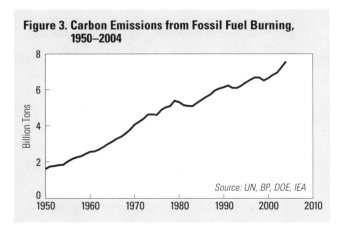

Source: UN, BP, DOE, IEA

Global Average Temperature and Carbon Emissions from Fossil Fuel Burning, 1950–2005, and Atmospheric Concentrations of Carbon Dioxide, 1960–2005

Year	Carbon Dioxide	Temperature	Emissions
	(parts per mill. by vol.)	(degrees Celsius)	(bill. tons of carbon)
1950	n.a.	13.87	1.63
1955	n.a.	13.89	2.04
1960	316.9	14.01	2.58
1965	320.0	13.90	3.14
1970	325.7	14.02	4.08
1975	331.2	13.94	4.62
1980	338.7	14.16	5.32
1981	339.9	14.22	5.16
1982	341.1	14.03	5.11
1983	342.8	14.25	5.10
1984	344.4	14.07	5.27
1985	345.9	14.03	5.43
1986	347.2	14.12	5.60
1987	348.9	14.27	5.73
1988	351.5	14.30	5.95
1989	352.9	14.19	6.07
1990	354.2	14.37	6.14
1991	355.6	14.32	6.23
1992	356.4	14.14	6.10
1993	357.0	14.14	6.10
1994	358.9	14.25	6.23
1995	360.9	14.37	6.40
1996	362.6	14.25	6.55
1997	363.8	14.40	6.68
1998	366.6	14.56	6.67
1999	368.3	14.33	6.51
2000	369.5	14.32	6.64
2001	371.0	14.47	6.82
2002	373.1	14.55	6.95
2003	375.6	14.52	7.25
2004	377.4	14.48	7.57
2005 (prel)	379.6	14.60	n.a.

Source: GISS, BP, IEA, CDIAC, DOE, and Scripps Inst. of Oceanography.

Weather-related Disasters Affect Millions

Zoë Chafe

Economic damages from weather-related disasters hit an unprecedented $204 billion in 2005, nearly doubling the previous record of $112 billion set in 1998 and reflecting the high number of disasters hitting built-up areas.[1] (See Figure 1.) Insured damages from weather-related disasters reached an estimated $92 billion, eclipsing all previous tallies since 1980 and more than doubling the losses in 2004.[2]

Hurricane Katrina, which alone caused an estimated $125 billion in damages to New Orleans and other areas of the southeastern United States, was one in a line of devastating hurricanes to hit Atlantic coasts in 2005—the most active hurricane season since 1851, the first year storms were tracked.[3] Three of the 10 strongest hurricanes ever recorded occurred in 2005.[4]

LINKS pp. 32, 42, 52

Climate studies have shown that warmer tropical ocean temperatures can cause more intense hurricanes, defined by higher wind speeds and increased precipitation.[5] Some scientists have linked climate change to an increase in sea surface temperatures, and at least one study has shown a recent marked increase in hurricane intensity.[6]

The number of people affected by weather-related disasters has jumped from an average of 97 million a year in the early 1980s to 260 million a year since 2001.[7] (See Figure 2.) At the same time, the number of people killed by weather-related natural disasters has decreased, on average. Some 13,670 people died because of weather-related disasters in 2005, slightly fewer than in the previous year.[8]

Though hurricanes and other dramatic storms easily attract media attention, with news sources continuously tracking wind speeds and storm paths, much of the devastation occurs on the sidelines of the storms, in the form of flash flooding and landslides.[9] This was the case in New Orleans, where much of the damage during Hurricane Katrina was in the form of flooding caused by a storm surge wave that was nine meters high, as well as in Central America, where torrential rains from Hurricane Stan caused 840 fatalities and left an additional 800 people missing after a landslide in Guatemala.[10]

Humans are not the only populations affected by weather-related disasters. In late 2005, many monkeys, toucans, and other wildlife died in Costa Rica's Corcovado National Park after three months of below-average temperatures and heavy rainfall.[11] Scientists studying the die-off are worried that climate change will produce more extreme weather events such as this one, leaving various animal populations vulnerable to stress and starvation.[12]

Salvano Briceño, director of the International Strategy for Disaster Reduction, warns that "environmental degradation, rapid urbanization, global warming, and the lack of institutional capacities are making millions of people more vulnerable to natural hazards."[13] Many of the people hardest hit by disasters cannot afford standard insurance policies and face extreme challenges as they try to rebuild their lives. A burgeoning microinsurance industry offers low-income populations the chance to purchase insurance against weather-related disasters and other risks through development banks and collective savings groups.[14]

In January 2005, 168 countries signed the Hyogo Framework for Action, which details the steps needed to reduce the impact of natural hazards on populations over the next 10 years.[15] Since then, 40 countries have adjusted their national policies to give disaster risk reduction greater priority.[16]

There are myriad examples of disaster planning paying off at the local level. When Hurricane Ivan hit Jamaica in 2004, for example, one community disaster response team issued early warnings by megaphone, then used risk maps and equipment assembled by the local Red Cross to evacuate all the area residents successfully.[17] In Cuba, where hurricane awareness is taught in schools, practice drills before each hurricane season and mandatory evacuation orders have helped save countless lives.[18] And in the Philippines, where on average 9 million people a year are affected by disasters, the Citizens' Disaster Response Center now helps communities build disaster management plans, provides emergency response, and assists with rehabilitation for disaster survivors.[19]

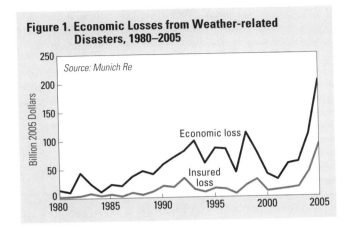

Figure 1. Economic Losses from Weather-related Disasters, 1980–2005

Source: Munich Re

Economic loss

Insured loss

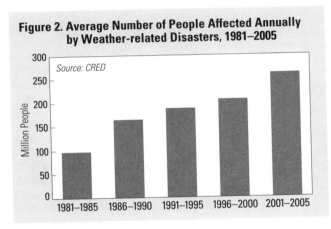

Figure 2. Average Number of People Affected Annually by Weather-related Disasters, 1981–2005

Source: CRED

Economic and Insured Losses from Weather-Related Disasters, 1980–2005

Year	Economic Loss	Insured Loss
	(billion 2005 dollars)	
1980	14.5	2.4
1981	10.6	2.7
1982	44.1	3.7
1983	24.8	7.8
1984	11.0	3.8
1985	23.7	6.2
1986	21.2	2.8
1987	37.3	8.8
1988	46.8	5.0
1989	41.1	10.5
1990	58.1	20.0
1991	69.8	17.7
1992	80.5	33.2
1993	98.6	13.5
1994	59.5	8.7
1995	85.6	15.2
1996	84.0	13.6
1997	42.9	5.4
1998	112.4	20.4
1999	78.2	30.1
2000	39.2	9.5
2001	29.9	11.6
2002	57.3	13.6
2003	60.8	16.2
2004	109.1	43.4
2005 (prel)	204.0	92.2

Source: Munich Re.

Hydropower Rebounds Slightly

Zijun Li

In 2004, the latest year for which data are available, worldwide hydroelectric power generation rose 5 percent after stagnating in 2003, reaching a total of 2,803 terawatt-hours (TWh).[1] (See Figure 1.) Hydroelectric power supplied 16.1 percent of global electricity in 2004, down from 18.5 percent in 1994.[2] Large-scale hydropower remains one of the lowest-cost technologies for generating electricity. Yet environmental constraints, difficulties resettling people who need to be moved when a dam is built, and the lack of appropriate sites have all hampered further development of large-scale dams in many countries.[3]

LINKS pp. 42, 58, 92

The top five producers—Canada (12 percent of the total), China (11.7 percent), Brazil (11.4 percent), the United States (9.4 percent), and the Russian Federation (6.3 percent)—generated nearly 51 percent of hydroelectric power in 2004.[4] (See Figure 2.) Several countries, including Brazil and Norway, obtain almost all their electricity from this one source.[5]

Expansion of hydropower is expected in the next few decades as large dams and reservoirs are completed in Asia.[6] China now ranks first globally in terms of total installed capacity.[7] Its total hydropower generation rose from 213 TWh in 1999 to 328 TWh in 2004, increasing at an average annual rate of 9.1 percent.[8] The Three Gorges Dam project on the Yangtze River generated 491 TWh of electricity in 2005; once completed in 2009, it is expected to produce 847 TWh annually.[9] The Chinese government, which views hydroelectric power as a renewable energy source that can be tapped on a large scale, aims for a total of 200–240 gigawatts of hydroelectric generating capacity by 2020.[10]

India's 15-gigawatt Nathpa Jhakri project on the Sutlej River in Himachal Pradesh became fully operational in 2004.[11] And in 2005, the World Bank approved financing for the $1.2-billion Nam Theun project in Laos.[12]

Total hydropower generation in the United States declined from a peak of 733 TWh in 1997 to 627 TWh in 2004, dropping at an average annual rate of 3.2 percent during that period due primarily to a record drought in the Pacific Northwest and the rising costs of project relicensing.[13] During the first half of 2005, 63 percent of U.S. hydroelectric output was supplied by just four states: Washington, California, Oregon, and New York.[14]

Scientists have recently drawn attention to the fact that hydroelectric dams produce significant amounts of carbon dioxide and methane—greenhouse gases closely connected to climate change.[15] Large amounts of carbon bound up in trees and other plants are released when a reservoir is initially flooded and the plants rot. And as plant matter settling on the reservoir's bottom decomposes without oxygen, it leads to a buildup of dissolved methane, which is released into the atmosphere when water passes through the dam's turbines.[16]

As the climate-related impacts of hydro projects are of increasing concern for the international community, starting in 2006 the Intergovernmental Panel on Climate Change will include emissions from artificially flooded regions in its proposed National Greenhouse Gas Inventories Programme, which calculates each country's carbon budget.[17]

To address various negative impacts of large dams, in 2000 the World Commission on Dams proposed 26 guidelines for dam-builders, covering such matters as respect for the environment, consultation with local people, proper plans for people who are displaced, and a cost-benefit analysis.[18] Despite rising concerns about the potential impacts of large dams on the environment and local poor people, in 2003 the World Bank announced that it would return to large-scale support for high-risk dams after having reduced its lending for such projects in the late 1990s.[19] So far World Bank–supported dams have led to the eviction of over 10 million people; the largest one, Nam Theun 2 in Laos, has already hurt the livelihoods of more than 100,000 people.[20]

Figure 1. World Hydroelectricity Use, 1965–2004

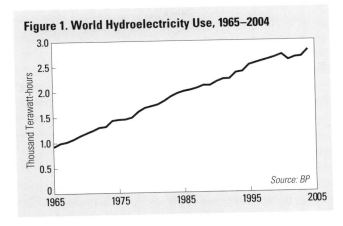

Source: BP

Figure 2. Hydroelectricity Use in Brazil, Canada, and China, 1965–2004

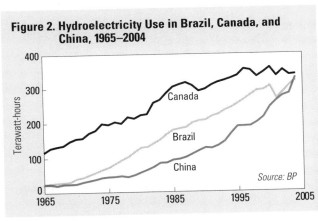

Source: BP

World Hydroelectricity Use, 1965–2004

Year	Use
	(terawatt-hours)
1965	927
1966	993
1967	1,018
1968	1,068
1969	1,133
1970	1,187
1971	1,237
1972	1,298
1973	1,312
1974	1,431
1975	1,449
1976	1,455
1977	1,490
1978	1,603
1979	1,678
1980	1,710
1981	1,743
1982	1,813
1983	1,900
1984	1,963
1985	2,004
1986	2,028
1987	2,063
1988	2,115
1989	2,113
1990	2,185
1991	2,237
1992	2,241
1993	2,363
1994	2,384
1995	2,513
1996	2,553
1997	2,589
1998	2,625
1999	2,664
2000	2,713
2001	2,606
2002	2,656
2003	2,669
2004	2,803

Source: BP.

Energy Productivity Gains Slow

Peter Stair

Global energy productivity, defined as gross world product divided by total energy use, increased by only 0.3 percent in 2004.[1] (See Figure 1.) In other words, energy use grew almost as rapidly as the global economy for two years in a row, as the pace of industrialization increased in many developing countries.[2]

This slowdown in energy productivity gains is a departure from a broader trend. In industrial economies, energy productivity has tended to increase at about 1 percent a year for decades.[3] And since 1970 global energy productivity has increased by 56 percent—about 1.3 percent each year.[4] If the world economy used energy today in the same way it did in 1970, it would require the energy equivalent of 11 additional Saudi Arabias—more than all the oil produced in the world.[5] Thus "saved energy" is arguably the world's leading energy source.

LINKS pp. 32, 52

The trend toward higher energy productivity is due to a combination of technology advances and structural economic shifts. Heavy industry is a smaller share of the economy, and the fastest-growing sectors—such as health care and education—require much less energy.[6] In addition, new technologies have improved the efficiency of everything from home appliances to industrial equipment.[7] Today, for example, compact fluorescent light bulbs provide as much light as incandescent bulbs but use one third as much electricity.[8]

In contrast, energy productivity often declines in newly industrializing countries. That is because traditional agriculture uses less energy per unit of output than most industries do.[9] Yet transitional countries today tend to introduce more-efficient technologies at an earlier stage, reducing the decline in energy productivity associated with industrialization.[10]

The oil crises of the 1970s dramatically increased the price of energy, providing a strong incentive to use energy more sparingly.[11] Between 1973 and 1983 the energy productivity of western industrial countries increased on average 2.2 percent a year, primarily due to the application of more-energy-efficient technologies.[12] After oil prices fell in the late 1980s, however, the improvements slowed.[13] For example, the fuel economy of U.S. cars nearly doubled between 1975 and 1988, but during the 1990s this progress stagnated as consumers demanded larger and more powerful cars.[14]

The collapse of the Soviet bloc closed some of the world's least efficient factories and power plants. In eastern Germany, many wasteful facilities were shut down or modernized, accelerating the improvement in Germany's energy productivity during the early 1990s.[15] (See Figure 2.) In Russia, progress was slowed by an economic recession, which contributed to a decline in energy productivity over the same period.[16]

Since the late 1990s, structural shifts in the economy have played an important role in boosting energy productivity. Low energy prices lessened the push for efficiency improvements.[17] Instead, growth in such high-tech and service sectors as computers, health care, and financial services propelled low-energy economic growth.[18] Between 1997 and 2003, productivity in western industrial countries increased about 1.8 percent a year.[19]

In recent years, developing countries' share of energy use has been growing—one factor in the slowdown in global energy productivity gains. For example, Thailand's proportion of world energy consumption doubled between 1990 and 2004, while its energy productivity fell by one third.[20] China is also becoming an increasingly significant user of energy, but as it industrialized in the 1980s, it also achieved huge increases in energy productivity.[21] Thanks largely to an aggressive program to encourage efficiency, between 1997 and 2000 Chinese energy productivity rose 13 percent annually.[22] Since 2000, however, it has declined almost 7 percent each year.[23]

Over the next decade, global energy productivity is likely to increase faster than it has recently. High energy prices will encourage efficiency improvements, while less-energy-intensive electronics and service industries will continue growing. Moreover, since energy conservation is usually the cheapest way to comply with greenhouse gas regulations, the Kyoto Protocol will push further gains in energy productivity.

Figure 1. GDP/Energy Ratio, 1970–2004

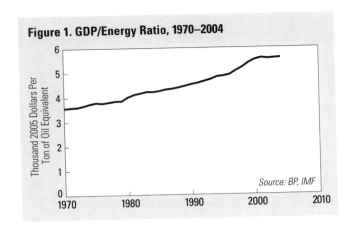

Source: BP, IMF

Figure 2. GDP/Energy Ratio for Selected Countries, 1980–2004

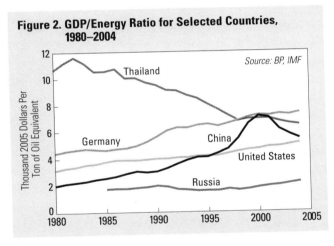

Source: BP, IMF

GDP/Energy Ratio, 1970–2004

Year	Ratio
	(thousand 2005 dollars per ton of oil equivalent)
1970	3.57
1971	3.59
1972	3.61
1973	3.66
1974	3.74
1975	3.79
1976	3.77
1977	3.80
1978	3.85
1979	3.85
1980	4.01
1981	4.11
1982	4.16
1983	4.22
1984	4.22
1985	4.26
1986	4.31
1987	4.34
1988	4.39
1989	4.45
1990	4.52
1991	4.57
1992	4.64
1993	4.71
1994	4.82
1995	4.85
1996	4.90
1997	5.06
1998	5.19
1999	5.37
2000	5.50
2001	5.57
2002	5.54
2003	5.55
2004	5.57

Source: BP, IMF.

Economic Trends

Photodisc

Roundwood ready for chipping

▶ Global Economy Grows Again

▶ Advertising Spending Sets Another Record

▶ Steel Output Up But Price Drops

▶ Aluminum Production Increases Steadily

▶ Roundwood Production Hits a New Peak

For data and analysis on economic trends, including world trade, gold mining, and tourism, go to www.worldwatch.org/vsonline.

Global Economy Grows Again

Erik Assadourian

Gross world product (GWP)—the sum of all finished goods and services produced globally—jumped 4.6 percent in 2005 to another record high of $59.6 trillion (in 2005 dollars).[1] (See Figure 1.) This total—based on the purchasing-power-parity valuation of country-based gross domestic product (GDP) measures—is roughly a percentage point higher than the annual average increase of 3.5 percent since 1970.[2] The market-rate GWP, which is based on actual monetary terms, grew 4.8 percent to $43.9 trillion.[3]

LINKS pp. 44, 48, 110

The U.S. economy, which makes up 21 percent of the GWP, grew by 3.8 percent.[4] One primary driver of this growth was high consumer spending propped up by the wealth generated by the booming housing market.[5] Total growth, however, was 0.5 percent less than earlier projections because of Hurricane Katrina.[6] This one natural disaster shrunk the GWP 0.1 percent.[7] Hurricanes—the severity of which could increase due to climate change—may continue to slow total GWP growth.[8]

The European Union, if counted as one entity, also makes up 21 percent of the GWP, and its economy grew by 2.7 percent in 2005.[9] This comparatively slow growth was mainly due to weak domestic demand.[10] China, which produced 14 percent of the GWP in 2005, grew at a rapid 7.8 percent.[11] Much of China's growth came from rapid expansion in its manufacturing exports, combined with a slowdown in its import growth.[12] Japan—7 percent of the global economy—grew by 2.6 percent, primarily due to strong consumer demand and a strengthening labor market.[13]

India accounts for 6 percent of the GWP and registered 7-percent growth in its economy.[14] This was driven mainly by growth in the service sector, such as information technology, and in industrial production.[15] Sub-Saharan Africa, home to 11.4 percent of the world's population, produced just 3 percent of the GWP.[16] Total economic output in the continent did increase 5.8 percent—aided by low inflation, rising oil prices, and fewer armed conflicts.[17]

Per capita GWP also grew in 2005, to $9,233 per person.[18] (See Figure 2.) Yet because world population increased by 74 million, per capita growth was only 3.4 percent.[19] Of course, as an aggregated sum per capita GWP does not capture the discrepancies across countries. In 2005, GDP per person in the United States was $41,701 and in Japan it was $31,466.[20] In China, though, GDP per person was $6,194, in India it was $3,335, and in sub-Saharan Africa it was only $2,075.[21]

The GDP per capita measure does not take into account unequal income distribution within countries. Thus in a country of relatively low income inequality, such as Sweden, GDP is more equitably distributed among the population. Compare this to the United States, where income inequality is 1.6 times that in Sweden.[22] The effects of this show up in other societal statistics: the probability of not surviving to age 60 in the United States is 64 percent higher than in Sweden and the population living on less than $11 a day is 2.2 times as high.[23]

One of the flaws of using GDP to measure economic progress is that it counts all economic activity as a positive addition, regardless of its societal worth. Moreover, some sectors of the economy, such as taking care of children and households, are excluded. The U.S. nongovernmental organization Redefining Progress has created an alternative measure—the genuine progress indicator (GPI)—that recalibrates the U.S. GDP by subtracting out pollution and other economic ills while adding in unmeasured benefits.[24] For 2002 (the most recent year with GPI data), Redefining Progress found the GPI to be $11,554 per person, less than a third of GDP per capita that year.[25] While GDP per capita grew by 79 percent between 1972 and 2002, the GPI grew just 1 percent.[26] (See Figure 3.)

Some people maintain that increased growth in the global economy is necessary to reduce poverty. A 2006 analysis, however, found that for every $100 worth of growth in GWP, only 60¢ contributed to reducing extreme poverty.[27] The recognition that growth in GWP often comes at the expense of the poor or the environment may lead policymakers and economists to give less priority to growth and more to better distribution.

Figure 1. Gross World Product, 1970–2005

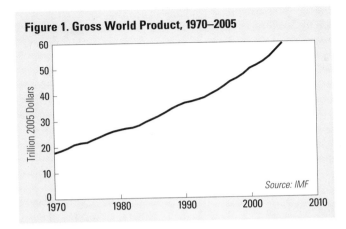

Source: IMF

Figure 2. Gross World Product Per Person, 1970–2005

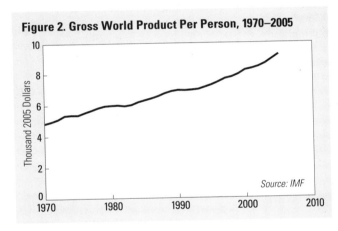

Source: IMF

Figure 3. GDP and GPI Per Person, United States, 1970–2002

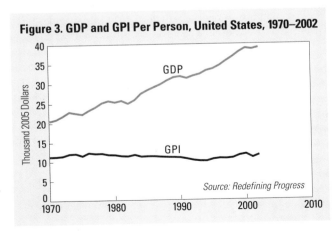

GDP

GPI

Source: Redefining Progress

Gross World Product, 1970–2005

Year	Total	Per Capita
	(trill. 2005 dollars)	(2005 dollars)
1970	17.9	4,829
1971	18.7	4,942
1972	19.7	5,094
1973	21.0	5,333
1974	21.6	5,369
1975	21.9	5,362
1976	23.0	5,525
1977	24.0	5,668
1978	25.1	5,830
1979	26.0	5,944
1980	26.6	5,977
1981	27.1	5,993
1982	27.4	5,946
1983	28.2	6,014
1984	29.5	6,190
1985	30.6	6,311
1986	31.7	6,427
1987	33.0	6,563
1988	34.5	6,747
1989	35.7	6,878
1990	36.7	6,946
1991	37.2	6,928
1992	37.9	6,955
1993	38.7	6,997
1994	40.1	7,136
1995	41.4	7,277
1996	43.1	7,464
1997	44.9	7,668
1998	46.1	7,766
1999	47.7	7,945
2000	49.9	8,212
2001	51.1	8,302
2002	52.5	8,433
2003	54.4	8,637
2004	57.0	8,933
2005 (prel)	59.6	9,233

Source: IMF.

Advertising Spending Sets Another Record   Erik Assadourian

Global advertising expenditures hit another record in 2005, jumping 2.4 percent to $570 billion.[1] (See Figure 1.) Nearly half of this total (48 percent) was spent in the United States, roughly the same as in 2004 but 4 percent less than in 2000.[2] One large source of new ad revenue in the United States was direct mail.

LINKS pp. 52, 120
Accounting for 21 percent of total U.S. advertising expenditures, direct mail grew 6 percent, to $56.6 billion and 41.5 billion pieces of mail.[3] This growth was in large part due to implementation of "Do Not Call" regulations in 2004, which restricted U.S. telemarketing practices.[4]

Of total ad spending, about $404 billion was directed toward major media, including television, newspapers, magazines, billboards, and the Internet—up 2.4 percent over 2004.[5] Television took the largest share, at 37 percent, but grew only 1.4 percent over the previous year.[6] Newspaper advertising, at 30 percent, also grew at 1.4 percent.[7] Magazine and radio advertising, which account for the third and fourth largest categories at 13 and 9 percent, stayed flat.[8] The fastest growth came from Internet advertising: though it accounted for only 5 percent of total major media advertising, it jumped 26 percent in 2005.[9]

Major media advertising grew slowly in North America and Europe—both at less than 1 percent.[10] It grew 2.6 percent in Asia, slightly more than the global average.[11] The majority of growth came in Latin America and in Africa and the Middle East, with these regions increasing almost 15 percent.[12] Combined, however, these regions account for only 10 percent of the global market.[13] (See Figure 2.)

While total advertising is now higher outside the United States than within it, per capita expenditures are still much smaller in the rest of the world.[14] (See Figure 3.) In 2005, total ad spending worldwide reached $88 per person. But in the United States it was more than 10 times this figure, at $933 per person.[15] Outside this country, $48 was spent per person on advertising.[16]

Advertising is a central means for businesses to increase their market share while also stimulating total consumer spending—often for products that are detrimental to consumer or environmental well-being. In 2004 the top 100 global marketers spent $96 billion on major media advertising.[17] Out of this, 24 percent was spent on automotive advertising, 17 percent went to ads on food and restaurants, soft drinks, and candy, and another 9 percent was spent promoting pharmaceuticals.[18]

With a growing obesity epidemic and tens of billions of dollars spent on advertising unhealthy foodstuffs each year, it is not surprising that many governments and consumer advocacy groups are trying to curb junk-food marketing, especially when it is targeted at children.[19] In response to threats by European Union commissioners to pass new advertising laws, soft drink companies agreed to voluntary restrictions on advertising, including not aiming ads at children under 12.[20]

France went a step further, requiring food marketers to include health messages in all their ads or pay a tax of 1.5 percent of their annual French advertising budgets to an institute that will promote healthy eating choices.[21]

The U.S. consumer group Center for Science in the Public Interest (CSPI) is using a different tactic to change marketing practices: litigation. In January 2006, CSPI filed a $2-billion suit against the entertainment company Viacom and the food company Kellogg (the thirty-seventh largest global advertiser in 2004) for engaging in unfair and deceptive marketing of "foods of poor nutritional quality" to children under 8.[22]

Another U.S. group, Commercial Alert, is working to ban direct-to-consumers (DTC) advertising by pharmaceutical companies—a practice that encourages consumers to request specific medications, even when the drugs are unnecessary or inappropriate.[23] According to a study by the Kaiser Family Foundation, every $1 spent on DTC advertising in 2000 produced an additional $4.20 in sales.[24] Desire to sell the drugs can lead pharmaceutical companies to bias DTC advertising or even exaggerate the frequency of conditions in order to stimulate sales of medications that are supposed to treat them.[25]

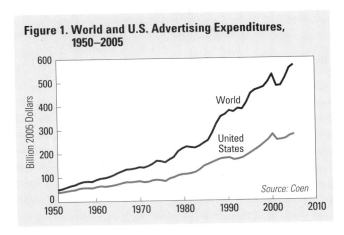

Figure 1. World and U.S. Advertising Expenditures, 1950–2005

Billion 2005 Dollars

World

United States

Source: Coen

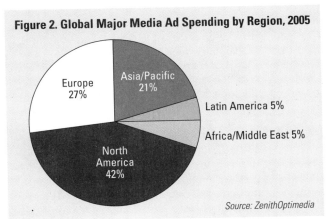

Figure 2. Global Major Media Ad Spending by Region, 2005

Europe 27%

Asia/Pacific 21%

Latin America 5%

Africa/Middle East 5%

North America 42%

Source: ZenithOptimedia

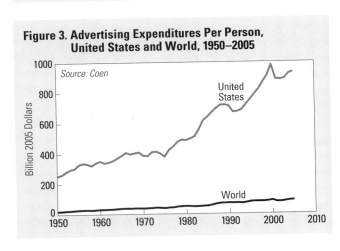

Figure 3. Advertising Expenditures Per Person, United States and World, 1950–2005

Source: Coen

Billion 2005 Dollars

United States

World

World and U.S. Advertising Expenditures, 1950–2005

Year	World	United States
	(billion 2005 dollars)	
1950	50	39
1955	78	55
1960	96	64
1965	126	76
1970	140	79
1971	146	80
1972	155	86
1973	168	88
1974	166	86
1975	161	82
1976	173	93
1977	183	98
1978	207	106
1979	219	110
1980	227	111
1981	224	114
1982	222	119
1983	230	130
1984	242	145
1985	253	152
1986	285	160
1987	324	168
1988	353	175
1989	362	177
1990	378	178
1991	373	170
1992	387	173
1993	384	178
1994	411	189
1995	450	200
1996	464	212
1997	470	224
1998	477	239
1999	498	254
2000	530	276
2001	481	252
2002	483	254
2003	515	258
2004	556	270
2005 (prel)	570	276

Source: Coen.

Steel Output Up But Price Drops

Yingling Liu

World crude steel production reached 1.13 billion tons in 2005, a 6.8-percent increase over 2004.[1] (See Figure 1.) Although global steel prices hit record highs in 2004, they dropped 21 percent in 2005 due to an inventory buildup in North America and Europe as well as China's expanding exports to the international market.[2] (See Figure 2.) The European export price, at a high of $582 per ton at the start of 2005, fell 32 percent to $393 by the end of the year, while the U.S. export price for iron and steel dropped 9 percent.[3] Meanwhile, global steel demand in 2005 stayed at about the same level as in 2004.[4]

China's expanding production, which accounted for almost a third of global supply, created a glut that increased exports and hurt prices in other regions.[5] Steel output there rose by 41 percent to 349 million tons.[6] (See Figure 3.) In 2005, China produced more steel than did Europe and North America combined.[7]

LINKS p. 52

Bloated production capacity, together with dampened domestic demand, forced more and more Chinese steel onto the export market.[8] China's exports leapt 185 percent in the first half of 2005, propelling the country from eighth to third position in the league of exporters, behind only Japan and Russia.[9] As imports again exceeded exports starting in July 2005, the annual increase in exports—at 20.5 million tons in 2005—was 44 percent.[10]

Steelmakers around the world have responded differently to the depressed steel prices. Pushed by the need to get better terms from suppliers, international steel companies are undergoing a succession of mergers.[11] Arcelor, a combined group of top companies in France, Spain, and Luxembourg, was toppled from its leading position earlier this year when Mittal, the London-based steel empire, bought the International Steel Group in the United States.[12] In the second half of 2005, a number of regional steelmakers in China also regrouped in order to integrate the highly fragmented market and improve market competitiveness.[13]

In order to prop up prices, steelmakers in Europe and China have cut output. Mittal Steel reduced its output by 1 million tons in the second half of 2005, while Arcelor cut production of flat stainless steel products in Europe by 15–20 percent since April.[14] Thyssenkrupp of Germany also cut its European stainless steel output by 20 percent, or 120,000 tons, in the July to September period.[15] China's major steel producers lowered output in the fourth quarter of 2005 to avoid further price drops.[16]

As the European and North American steel industry consolidated, production was brought in line with demand.[17] North American steel output fell by 7 million tons in 2005, or 5.3 percent, to 127 million tons, and European production decreased by 3.6 percent to 186.5 million tons.[18] As a result, the free fall in steel prices was halted at the end of 2005.

Expanding steel production has strained the supply of raw materials, creating shortages and driving up prices of feedstock and fuel in the global market. The price of coking coal, which fuels blast furnaces, climbed from nearly $69 to $86 a ton by September 2005, while iron ore jumped from $38 per ton at the end of 2004 to $65 by the end of 2005, a record 72-percent rise.[19]

The other feedstock—scrap steel—has seen a significant boom in 2005. Recycled scrap steel is mostly used in electric arc furnaces, which produce about one third of global steel (350 million tons) by producing virtually a ton of steel out of every ton of scrap.[20] About 15 percent of China's steel output (roughly 40 million tons) is produced by this scrap-hungry route, and the country needs to import about 10 million tons of scrap steel a year to supplement its local supply.[21] This makes China the world's second-largest importer of scrap steel, after only Turkey.[22]

The steel industry in North America has seen a rise in steel scrap recycling. In the United States alone, an estimated 67 million tons of steel were recycled in 2005.[23] The total value of U.S. domestic purchases and exports of steel scrap was estimated to be $12.6 billion in 2005, up about 32 percent from the figure in 2004.[24] As recycling becomes more profitable and convenient, the world will see more steel being created out of old scrap.

Figure 1. World Steel Production, 1950–2005

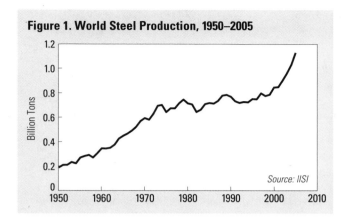

Source: IISI

Figure 2. Global Steel Prices, March 2002–December 2005

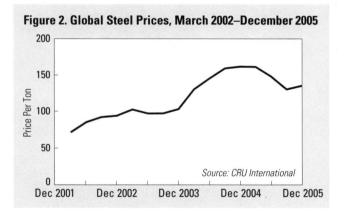

Source: CRU International

Figure 3. Top Five Steel-Producing Countries, 1994–2005

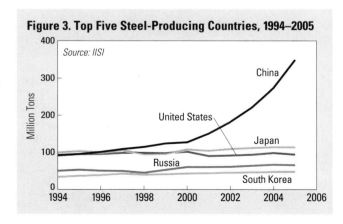

Source: IISI

World Steel Production, 1950–2005

Year	Production
	(million tons)
1950	190
1955	271
1960	347
1965	451
1970	595
1971	582
1972	631
1973	696
1974	703
1975	644
1976	675
1977	675
1978	717
1979	747
1980	716
1981	707
1982	645
1983	664
1984	710
1985	719
1986	714
1987	736
1988	780
1989	786
1990	771
1991	734
1992	720
1993	728
1994	725
1995	752
1996	750
1997	799
1998	777
1999	788
2000	847
2001	850
2002	904
2003	962
2004	1,057
2005 (prel)	1,129

Source: IISI.

Aluminum Production Increases Steadily

Andrew Wilkins

In 2005, production of world primary aluminum, which is made from raw ore, climbed almost 5 percent to a record 31.2 million tons.[1] Production of secondary, or recycled, aluminum increased slightly to 7.6 million tons in 2004, the last year with data.[2] (See Figure 1.) Annual production of primary aluminum has increased more than 20-fold since 1950, and world per capita production has swelled from 0.6 to 4.8 kilograms.[3]

Primary aluminum is smelted from alumina (aluminum oxide), which is refined from bauxite ore, a reddish clay mined most heavily in Australia, Guinea, Brazil, and China.[4] In general, four tons of bauxite yield two tons of alumina, which in turn yield one ton of aluminum metal.[5] In 2004, an estimated 156 million tons of bauxite were mined worldwide, the vast majority of which was converted into aluminum.[6] (The remainder went to non-metallurgical uses, such as abrasives, chemicals, and refractories.)[7]

LINKS pp. 32, 42, 46

Aluminum is strong, light, easily worked, corrosion-resistant, and a good conductor of heat and electricity, and it can be alloyed easily with other metals.[8] It is used mainly in the transportation, packaging, construction, electrical, consumer goods, and machinery sectors.[9]

Since 2002, China has been the number one producer of primary aluminum in the world, smelting more than 20 percent of the total output in 2004.[10] As recently as 2000, the United States was at the top, but it has fallen behind China, Russia, and Canada. These four countries account for more than 50 percent of the world's primary aluminum output.[11]

China is now also the leading consumer of primary aluminum, accounting for 20 percent of world use in 2004.[12] (See Figure 2.) On a per capita basis, however, the Chinese only used 4.6 kilograms, compared with Americans at 29.5 kilograms and Europeans at 19.7 kilograms apiece.[13] People in most developing countries use on average 1–2 kilograms each year.[14]

Secondary production of aluminum has fallen from a high of 8.2 million tons in 2000.[15] One important characteristic of aluminum is that it can be recycled indefinitely without diminishing in quality. The recycling process uses only 5–8 percent as much energy as primary production does and about 10 percent as much capital equipment.[16] The United States is still the leader in secondary aluminum production, recycling nearly 40 percent of the world total—some 3 million tons—in 2004.[17] Roughly 60 percent of this came from manufacturing scrap, with the rest from discarded, post-consumer products such as used beverage cans.[18]

Aluminum is the third most plentiful element on Earth's crust.[19] Minuscule amounts of aluminum were smelted from bauxite as early as 1825, and production increased modestly through the middle of that century. During this time, aluminum was considered a precious metal and initially sold at prices well above platinum and gold. As production techniques improved, prices fell. Then in 1886 the basis for modern aluminum smelting, the Hall-Héroult electrolytic process, was introduced and production soared.[20]

The Hall-Héroult process uses a tremendous amount of electricity, making aluminum production one of the world's most energy-intensive industries and a significant source of greenhouse gas emissions. Producing one ton of aluminum requires on average 15,270 kilowatt-hours of electricity, enough to run an average American household for nearly 18 months.[21] Primary smelters used 421 billion kilowatt-hours of electricity in 2003, equal to nearly 3 percent of world electricity consumption.[22] Aluminum production also emits a significant amount of perfluorocarbons, gases with extremely high climate change potential.[23]

While modern smelters use about 40 percent less electricity than their 1950s' counterparts did, overall electricity use in the industry is on the rise.[24] (See Figure 3.) A major switch toward secondary aluminum production could slow this trend but is unlikely to reverse it.[25] In 2003, only about 44 percent of aluminum beverage cans in the United States were recycled—Americans tossed the rest, about 2.5 million tons of aluminum, into landfills.[26] If this metal had been recycled, it could have saved 36.7 billion kilowatt-hours of electricity, enough to power 3.5 million U.S. households for a whole year.[27]

Figure 1. World Aluminum Production, 1950–2005

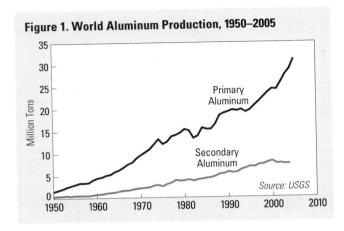

Figure 2. Primary Aluminum Use by World Area, 2004

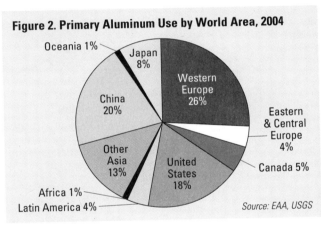

Figure 3. World Electricity Use in Primary Aluminum Production, 1951–2004

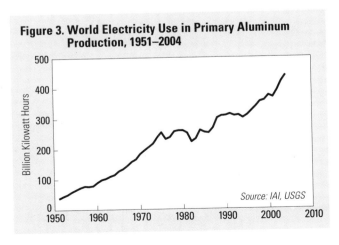

World Aluminum Production, 1950–2005

Year	Primary	Secondary
	(million tons)	
1950	1.5	0.4
1955	3.1	0.6
1960	4.5	0.9
1965	6.3	1.5
1970	9.7	2.2
1971	10.3	2.3
1972	11.0	2.4
1973	12.1	2.8
1974	13.2	2.9
1975	12.1	2.6
1976	12.6	3.1
1977	13.8	3.4
1978	14.1	4.0
1979	14.6	3.8
1980	15.4	3.9
1981	15.1	4.1
1982	13.4	3.8
1983	13.9	4.1
1984	15.7	4.2
1985	15.4	4.4
1986	15.4	4.5
1987	16.5	4.8
1988	18.5	5.3
1989	19.0	5.4
1990	19.3	5.8
1991	19.7	5.6
1992	19.5	5.7
1993	19.8	6.3
1994	19.2	6.6
1995	19.7	7.0
1996	20.7	6.9
1997	21.6	7.4
1998	22.6	7.5
1999	23.6	7.9
2000	24.4	8.2
2001	24.3	7.6
2002	25.9	7.7
2003	27.7	7.5
2004	29.8	7.6
2005 (prel)	31.2	n.a.

Source: USGS.

Roundwood Production Hits a New Peak

Zijun Li

Global roundwood production reached a new peak of 3,402 million cubic meters in 2004, the last year for which data are available.[1] (See Figure 1.) Just five countries—the United States, India, China, Brazil, and Canada—produced 44 percent of the global roundwood harvest.[2] Fifty-nine percent of the total was harvested in developing nations, a slight drop from 61 percent in 1999.[3] (See Figure 2.)

p. 102

The wood harvested in 2004 was used about equally for industrial purposes and for heating and cooking. Fuelwood and charcoal accounted for 52 percent of the total harvest, while the other 48 percent was made into lumber, panels for construction, or pulp for paper.[4]

Developing countries accounted for 90 percent of the wood cut for fuel but only 26 percent of the industrial wood harvest.[5] In contrast, industrial countries—home to less than 20 percent of global population—cut 74 percent of the wood used for industrial purposes.[6] The United States, the largest consumer in this category, used 450 million cubic meters of industrial roundwood in 2004, the most since 1999.[7]

While industrial countries have harvested relatively consistent amounts of industrial roundwood each year since 1970, developing countries have doubled their output.[8] As manufacturing activity and the use of industrial wood products keep growing in developing regions, future wood production will be driven significantly by developments in China, other countries in Asia and the Pacific region, Latin America, South Africa, the Russian Federation, and eastern Europe.[9]

China's imports of industrial roundwood reached 27 million cubic meters in 2004, making this nation the largest importer of industrial roundwood.[10] In part, this is due to decreasing Chinese production. China's output of industrial roundwood dropped from 107 million cubic meters in 1998 to 95 million cubic meters in 2004 thanks to diminished forest cover and limits on domestic logging.[11]

At the same time, the Russian forest sector is beginning to recover from collapse following the breakup of the Soviet Union. Its industrial roundwood output nearly doubled in eight years—from 73 million cubic meters in 1996 to 134 million cubic meters in 2004—making it the third largest industrial-roundwood producer, after the United States and Canada.[12] Russia's roundwood exports reached 41.8 million cubic meters in 2004, accounting for 30 percent of its reported roundwood harvest as well as 30 percent of global roundwood exports.[13]

Production of sawlogs and veneer logs—roundwood that will be sawn or chipped lengthways to make sawnwood or railway ties or used for the production of veneer—reached 960 million cubic meters in 2004.[14] This currently accounts for the largest share of industrial roundwood output, at 59 percent, dropping slightly from 63 percent in 1961.[15] (See Figure 3.) Production of pulpwood, which is used to make pulp, particleboard, or fireboard, has increased from 447 million cubic meters in 1998 to 509 million cubic meters in 2004—31 percent of the total industrial roundwood output.[16] Output of other industrial roundwood that is used for tanning, distillation, match blocks, and piling dropped from a peak of 230 million cubic meters in 1989 to 155 million cubic meters in 2004.[17]

Substantial volumes of roundwood are removed illegally from important supply regions. Eastern Russia, Indonesia, Brazil, and West Africa are among the regions with the largest problems.[18] Areas with larger exports also have high amounts of illegal logging. A study by the American Forest and Paper Association in 2004 estimated that 15–20 percent of the harvest might be defined as illegal in the Russian Federation as a whole.[19] It also showed that about 8 percent of the world's roundwood harvest is from suspicious sources.[20]

As the international community has increasingly recognized this problem, governments have made commitments to combat illegal logging and trade.[21] Indonesia, for instance, has signed memoranda of understanding or other treaties with major tropical log-importing countries in an attempt to control illegal logging and trade of timber products.[22] It also reduced its logging quota from 12 million cubic meters in 2002 to 5.74 million cubic meters in 2004.[23]

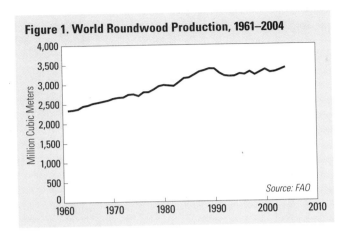

Figure 1. World Roundwood Production, 1961–2004

Source: FAO

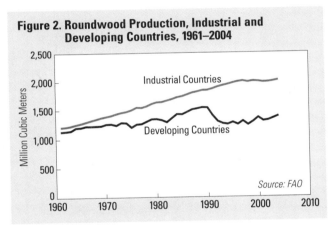

Figure 2. Roundwood Production, Industrial and Developing Countries, 1961–2004

Industrial Countries

Developing Countries

Source: FAO

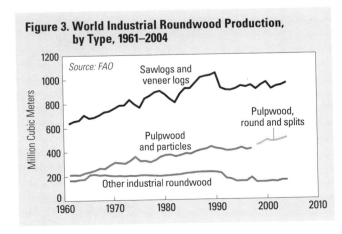

Figure 3. World Industrial Roundwood Production, by Type, 1961–2004

Source: FAO

Sawlogs and veneer logs

Pulpwood, round and splits

Pulpwood and particles

Other industrial roundwood

World Roundwood Production, 1961–2004

Year	Production
	(million cubic meters)
1961	2,342
1962	2,353
1963	2,377
1964	2,451
1965	2,475
1966	2,520
1967	2,543
1968	2,571
1969	2,598
1970	2,644
1971	2,666
1972	2,672
1973	2,740
1974	2,749
1975	2,705
1976	2,799
1977	2,801
1978	2,868
1979	2,950
1980	2,978
1981	2,965
1982	2,954
1983	3,048
1984	3,148
1985	3,162
1986	3,234
1987	3,309
1988	3,343
1989	3,386
1990	3,382
1991	3,272
1992	3,202
1993	3,188
1994	3,193
1995	3,251
1996	3,234
1997	3,304
1998	3,224
1999	3,292
2000	3,357
2001	3,284
2002	3,301
2003	3,348
2004	3,402

Source: FAO.

Transportation and Communication Trends

Refurbished bikes ready for sale at the Salvadoran Center for Appropriate Technology, El Salvador

▶ Vehicle Production Continues to Expand

▶ Bicycle Production Up

▶ Air Travel Takes Off Again

▶ Internet and Cell Phone Use Soar

For data and analysis on transportation and communications trends, including passenger rail travel, go to www.worldwatch.org/vsonline.

Vehicle Production Continues to Expand

Michael Renner

Global passenger car production grew 3.2 percent in 2005 to 45.6 million units, according to estimates by London-based Global Insight.[1] (See Figure 1.) Continuing the relentless growth in output, this sets a new record. In addition to traditional passenger cars, production of sports utility vehicles and other "light trucks" reached 18.5 million units, for a combined total of 64.1 million.[2]

According to Ward's Communications, there were 603 million passenger cars on the world's roads in 2004, the last year with such data, plus another 223 million commercial vehicles.[3] With a combined population of about 850 million, the United States, Canada, Japan, and Western Europe had 552 million vehicles.[4] China and India, with 2.3 billion people, had just 39.2 million—only 5 percent of the global total.[5] (See Figure 2.)

LINKS pp. 32, 42, 52

The top 10 producers are the United States, Japan, Germany, France, South Korea, Spain, China, Brazil, the United Kingdom, and Canada.[6] In the United States, passenger car production is now less than half its peak in the 1970s.[7] Yet light truck production has surged since the early 1980s, keeping that country on top as the leading vehicle producer.

China and, to a lesser extent, India are rapidly increasing their output of passenger cars. China's has risen from fewer than 100,000 in 1991 to 2.3 million in 2004, a 28-fold expansion that mirrors what happened in Japan in the 1960s.[8] With 940,000 units in 2004, India is now the twelfth-largest producer, although its pace is dwarfed by China's.[9]

Production is dominated by a handful of companies. The top 5—General Motors (GM), Ford, Toyota, Volkswagen, and DaimlerChrysler—produced 51 percent of all passenger cars and light trucks in 2004, while the top 10 accounted for 75 percent.[10]

Profits for the "Big Three" U.S. companies—GM, Ford, and DaimlerChrysler—are mostly derived from light trucks, which accounted for 47 percent of their output in 2004, compared with just 17 percent for all other companies.[11] But rising gasoline prices are lessening the appeal of these vehicles.[12] Since 2000, the Big Three have implemented or announced plant closings and layoffs that affect 140,000 jobs.[13]

Toyota, in contrast, is expanding production and may soon overtake General Motors as the top producer.[14] Toyota is the unquestioned leader in developing hybrid gasoline-electric vehicles, having sold 425,000 such cars since 1997.[15] Hybrid sales in the United States, dominated by Toyota, rose from 9,350 in 2000 to about 187,000 in 2005.[16] The company is planning to manufacture 400,000 hybrids in 2006 and hopes to sell 1 million of them by 2010.[17]

Hybrids are often seen as shorthand for greater fuel efficiency, yet Toyota and other car manufacturers are increasingly introducing "muscle hybrids," using the technology to boost acceleration and horsepower instead of improving fuel economy.[18] Under such circumstances, the rise of hybrids will offer incremental improvements at best—insufficient to temper the voracious appetite for gasoline or reduce the growing environmental impacts associated with expanding fleets.

Government and corporate actions to address fuel efficiency remain inadequate. Out of concern over rising prices and supply shortages, China is planning to introduce steep taxes that might add as much as 27 percent to the purchase price of a vehicle.[19] This follows the introduction of relatively stringent fuel economy standards in July 2005.[20]

Nowhere are such measures more needed than in the United States. Americans used 44 percent of the 20.2 million barrels per day of motor gasoline burned up worldwide in 2002, the most recent year with data.[21] (See Figure 3.) But U.S. fuel economy standards for passenger cars are no higher today than in 1985.[22] Only modest improvements, from 21.2 miles per gallon to 24 miles by 2011, are planned for light trucks.[23] The largest of these would be exempt, and manufacturers may be able to reclassify vehicles in ways that would limit actual efficiency gains. The auto industry is opposed to California's efforts to impose strict carbon emission rules that, by implication, would require sharp increases in fuel efficiency.[24]

Figure 1. World Passenger Vehicle Production, 1950–2005

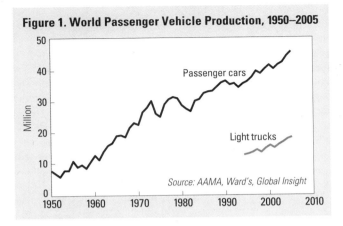

Passenger cars

Light trucks

Source: AAMA, Ward's, Global Insight

Figure 2. Share of World Motor Vehicle Fleet, Selected Countries and Regions, 2004

Western Europe 27.4%

Japan 8.7%

China and India 4.7%

Russia and Eastern Europe 8.7%

United States 27.6%

Rest of the World 22.9%

Source: Ward's

Figure 3. Share of Motor Gasoline Consumption, Selected Countries and Regions, 2002

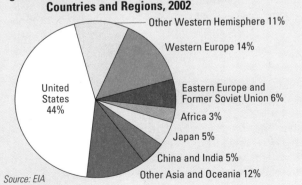

Other Western Hemisphere 11%

Western Europe 14%

Eastern Europe and Former Soviet Union 6%

Africa 3%

Japan 5%

China and India 5%

Other Asia and Oceania 12%

United States 44%

Source: EIA

World Passenger Vehicle Production, 1950–2005

Year	Passenger Cars	Light Trucks
	(million)	
1950	8.0	
1955	11.0	
1960	12.8	
1965	19.0	
1970	22.5	
1975	25.0	
1976	28.9	
1977	30.5	
1978	31.2	
1979	30.8	
1980	28.6	
1981	27.5	
1982	26.7	
1983	30.0	
1984	30.5	
1985	32.4	
1986	32.9	
1987	33.1	
1988	34.4	
1989	35.7	
1990	36.3	
1991	35.1	
1992	35.5	
1993	34.2	
1994	35.4	12.9
1995	36.1	13.2
1996	37.4	13.7
1997	39.4	14.5
1998	38.6	13.7
1999	40.1	15.0
2000	41.3	15.9
2001	40.1	15.1
2002	41.4	16.2
2003	42.2	17.0
2004	44.2	18.0
2005 (prel)	45.6	18.5

Source: American Automobile Manufacturers Association, Ward's, and Global Insight.

Bicycle Production Up

Gary Gardner

World bicycle production increased by nearly 9 percent to some 101 million units in 2003, the last year for which global data are available.[1] (See Figure 1.) The upswing marks the second consecutive year of healthy increase in production.

China's share of global bicycle production hovered at roughly 40 percent through most of the 1990s but has surged since 2000: some 58 percent of the bicycles produced worldwide in 2003 were made in China.[2] (See Figure 2.) The country's total output of 58 million bikes was nearly five times greater than the next largest producer, India, which built 12 million in 2003.[3] Chinese gains have come at the expense of other once-powerful producers, including Taiwan, Japan, the United States, Germany, and France, all of which have seen declines in market share and total production over the last decade.

LINKS pp. 118, 120

Much of the growth in Chinese bicycle production—and in Vietnamese production as well—is being fueled by overseas demand. The two Asian nations exported more than four times as many bikes to Europe in 2004 as in 2000.[4] In 2005, this prompted the European Union to impose hefty anti-dumping tariffs of 48.5 percent on Chinese bicycles and 34.5 percent on Vietnamese bikes.[5] The impact could be substantial: European tariffs in the early 1990s cut Chinese exports to the region from more than 2 million to roughly 50,000 in just two years.[6] Meanwhile, Mexican tariffs of 144 percent have shut that country's door to Chinese exports almost entirely.[7] The stiff tax was first imposed in 1994 after China had captured a quarter of the Mexican market and was renewed in 2005.[8]

One of the brightest signs on the bicycle landscape is the growth in output of electric bicycles, which have electric motors to make pedaling easier. Global sales reached 10.5 million units in 2005, a 79-fold increase over a decade.[9] (See Figure 3.) The rapid rise is driven largely by Chinese sales, which accounted for just 15 percent of global sales in 1996 but 95 percent in 2005.[10] Roughly one of every six bicycles bought in China in 2005 was an electric model.[11]

Several factors can encourage the nonpolluting, healthy contributions that bicycles provide as an integral part of a sustainable transportation mix. A 2005 study of 34 American cities, for example, found that commuting by foot and by bicycle is closely correlated with the amount of green space in a city.[12] The results suggested that cities can encourage more bicycle use by building more parks and greenways.

Astute marketing, government policies, and other factors can lift bicycle use to higher levels. Bike sales in Ghana, for example, stand at about 30 per 1,000 population, some 50 percent higher than in China, a wealthier country.[13] This is due in part to biking groups' and manufacturers' efforts to create high-quality, affordable bikes tailored to the needs of Ghanaians. Similarly, the bike ownership rate in the Netherlands is roughly 50 percent higher than the level Dutch incomes would suggest, in part because the government invests in safe cycling infrastructure.[14]

Bike sharing schemes have evolved over the years as novel ways to make cycling an easy and inexpensive option. In Lyon, France, a new subscription-based program was introduced in 2005 with 1,500 bicycles and about 100 computerized bike racks.[15] Subscribers pay 5 euros (about $6) a year, which gives them a computerized card to unlock bikes at any rack.[16] They also pay 1 euro per hour for usage—but the first half hour is free, and 90 percent of trips fall within this discounted window.[17] The program attracted some 15,000 subscribers in its first three months, and each bike is used 6.5 times per day, on average, for nearly 10,000 daily trips.[18]

A different approach is used in Arcata, California, where a "bike library" provides bikes to residents for six months for a $20 refundable fee.[19] In its first two years, the program registered 1,500 bike "checkouts" in the town of 16,000 people.[20] Local bike stores report that the library actually spurs sales because it raises interest in cycling and users want to move up to more-sophisticated bikes than the ones on loan.[21]

Figure 1. World Bicycle Production, 1950–2003

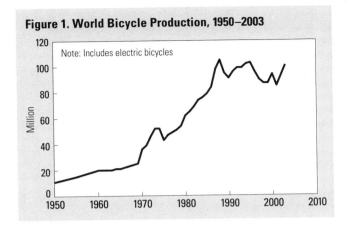

Note: Includes electric bicycles

Figure 2. Chinese Share of Global Bicycle Production, 1990–2003

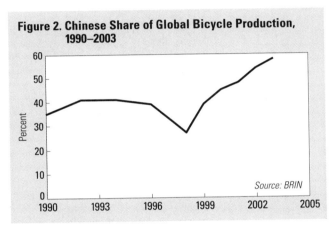

Source: BRIN

Figure 3. Sales of Electric Bicycles, 1996–2005

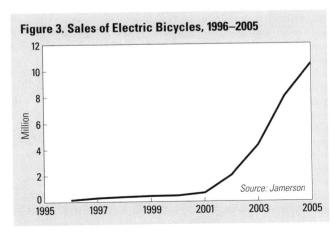

Source: Jamerson

World Bicycle Production, 1950–2003

Year	Production
	(million)
1950	11
1955	15
1960	20
1965	21
1970	36
1971	39
1972	46
1973	52
1974	52
1975	43
1976	47
1977	49
1978	51
1979	54
1980	62
1981	65
1982	69
1983	74
1984	76
1985	79
1986	84
1987	98
1988	105
1989	95
1990	91
1991	96
1992	99
1993	99
1994	102
1995	103
1996	96
1997	90
1998	87
1999	87
2000	94
2001	85
2002	93
2003	101

Source: Bicycle Retailer and Industry News and United Nations.

Air Travel Takes Off Again

Zoë Chafe

Air travel resumed its quick rise in 2004, the latest year for which data are available from the International Civil Aviation Organization (ICAO). The volume of passengers and the number of passenger-kilometers increased at rates not seen since the 1970s.[1] (See Figures 1 and 2.) Passenger numbers reached an estimated 1.9 billion in 2004, up 11 percent from 2003, and were projected to top 2 billion in 2005.[2] In the past 20 years, the flow of air travelers has more than doubled despite setbacks including terrorism, the war in Iraq, and health scares, such as severe acute respiratory syndrome.[3]

LINKS pp. 32, 42

The number of passenger-kilometers climbed an estimated 14 percent to 3.4 trillion in 2004, spurred by strong economies in the Middle East and cheaper flying options.[4] This figure is projected to soar to 4.2 trillion by 2007, triple the distance traveled 25 years earlier.[5]

Despite rising passenger numbers, only about 5 percent of the world's population has ever flown.[6] The average American flies twice as far each year as the average European, who in turn logs 10 times the distance of the average Asian.[7] Between 2005 and 2007, air traffic is expected to grow most quickly in the Middle East and Asia/Pacific regions.[8]

Because planes emit carbon dioxide (CO_2), water vapor, and nitrogen oxides high in the atmosphere, their emissions contribute to climate change at about three times the rate of similar emissions from cars.[9] The water vapor can form white "contrails" that, along with nitrogen oxide and CO_2 emissions, cause heat to reflect back to Earth.[10]

The two main aviation manufacturers, Airbus and Boeing, reported receiving 2,057 new plane orders during 2005.[11] Each month, Airbus produces an average of eight wide-body jets and 30 single-aisle planes.[12] Banking on rising demand for long-haul flights, Boeing introduced a new 300-passenger plane that flew more than 22,000 kilometers nonstop on its maiden flight in October 2005.[13] Such long flights may eliminate fuel-intensive takeoffs and landings, but they often cause more heat-trapping contrails than planes flying at lower altitudes.[14] One new report calls on manufacturers to focus on enabling planes to fly more efficiently at lower altitudes.[15]

Major improvements in aircraft operation over the past 40 years have made planes up to 70 percent more efficient, cleaner, and quieter.[16] And in 2005, Brazilian aviation company Neiva introduced the first plane designed to run on ethanol, a renewable biofuel that is cheaper and less polluting than kerosene, the most common jet fuel.[17]

Despite these advances, "the growth of air transport is currently outpacing progress in aviation environmental protection," concedes ICAO Council President Assad Kotaite.[18] A new ICAO environmental campaign, The Greening of Flight, will explore whether plane emissions could be integrated into emissions trading schemes and whether emissions taxes would address local air quality problems.[19]

In the United Kingdom, airlines, airports, and plane manufacturers launched a "sustainable aviation" campaign in 2005, pledging to improve fuel efficiency, reduce pollution, and include aircraft emissions in the European Union's emissions trading scheme.[20] Critics would prefer taxes on jet fuel and flights, noting that a threefold increase in air travel in the next 30 years would cancel out any improvements.[21] Another proposal suggests printing climate change warnings on air tickets to notify travelers of the impacts of their flights.[22]

The Kyoto Protocol, which sets targets for industrial countries to limit greenhouse gas emissions, does not cover releases from air travel, even though planes emit 600 million tons of CO_2 each year.[23] But industrial countries are expected to work with ICAO to reduce aircraft emissions. ICAO has identified global cooperation and consensus as the most effective way to minimize aviation's impact on the global environment.[24]

Individual travelers, as well as businesses, are now able to offset the CO_2 emitted during their flights by purchasing carbon credits from a variety of programs that use the funds to support renewable energy projects, such as solar cooking in India, or responsible waste management in Brazil.[25]

Figure 1. World Air Travel by Distance, 1950–2004

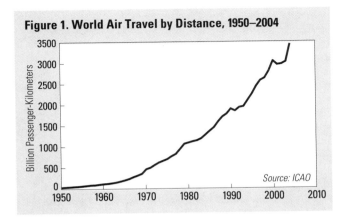

Source: ICAO

Figure 2. World Passenger Air Travel by Volume, 1950–2004

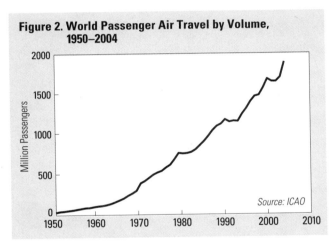

Source: ICAO

World Air Travel by Distance and Passenger Volume, 1950–2004

Year	Distance	Passengers
	(billion passenger-kilometers)	(million)
1950	28	31
1955	61	68
1960	109	106
1965	198	177
1970	460	383
1975	697	534
1976	764	576
1977	818	610
1978	936	679
1979	1,060	754
1980	1,089	748
1981	1,119	752
1982	1,142	766
1983	1,190	798
1984	1,278	848
1985	1,367	899
1986	1,452	960
1987	1,589	1,028
1988	1,705	1,082
1989	1,774	1,109
1990	1,894	1,165
1991	1,845	1,135
1992	1,929	1,146
1993	1,949	1,142
1994	2,100	1,233
1995	2,248	1,304
1996	2,432	1,391
1997	2,573	1,457
1998	2,628	1,471
1999	2,798	1,562
2000	3,038	1,672
2001	2,950	1,640
2002	2,965	1,639
2003	3,019	1,691
2004	3,442	1,887

Source: ICAO.

Internet and Cell Phone Use Soar

Zoë Chafe

The number of cell or mobile phone subscribers worldwide has tripled in five years, reaching 1.75 billion in 2004 according to the latest data from the International Telecommunications Union.[1] (See Figure 1.) Preliminary data from Computer Industry Almanac, Inc. indicate that the number of cell phone subscribers reached 2.1 billion in 2005.[2] While 15 countries have more than 90 cell phone subscribers per 100 residents, more than 40 countries still have fewer than 10 per 100 inhabitants.[3]

The research firm Gartner estimated that cell phone sales reached 816.6 million in 2005, a 21-percent increase over 2004, with several brands focused on selling simple low-cost phones to customers in developing countries.[4] China remains the largest cell phone market in the world, with 19 percent of the world's cell phone subscribers (394 million) and 92 million phones purchases in 2004.[5]

The number of Internet users worldwide topped 1 billion for the first time in 2005.[6] Germany has more than twice as many Internet users as the whole African continent, which with 22 million users has a meager 3 percent Internet penetration rate.[7] Some 19 countries count more than 50 Internet users per 100 people, while in another 100 countries, less than 10 percent of the population logs on.[8] (See Figure 2.)

The prevalence of Internet host computers has skyrocketed from less than 1 million 15 years ago to 395 million in 2005.[9] (See Figure 3.) But as the Internet continues its global spread, its international accessibility may falter. Pushing back at the U.S. government's continued involvement in the administration of domain names and the use of English as the default language in the development of the Internet, China has introduced three Chinese character suffixes that substitute for dot-com and are inaccessible to users outside the country, increasing fears about that government's ability to reroute and censor Web sites.[10]

Innovative programs are proving that the world's poorest people benefit from access to communications technologies. In Bangladesh, the Grameen Bank has provided small loans to 195,000 rural entrepreneurs who purchase cell phones to earn money selling calls to neighbors, enabling people to talk to far-flung family members or business contacts.[11] And in South Africa, where less than half the population has access to banking services, a major bank has linked ATM cards to prepaid cell phone accounts, allowing rural cell phone subscribers to deposit and transfer funds quickly and securely.[12]

With the average life span of a cell phone estimated at only 14 months, the issue of cell phone disposal has become critical.[13] In the United Kingdom, only about 10–15 percent of phones are presently recycled.[14] In the United States there may already be 500 million obsolete cell phones, and only 2.3 percent of Americans recycled their cell phones in 2004.[15]

Refurbishing and recycling used electronics means there is less need to mine for new materials—including lead, copper, mercury, and coltan—to produce computers and cell phones.[16] Coltan, an ore found almost exclusively in the war-torn Democratic Republic of the Congo, is used in computer and cell phone capacitors. Coltan mining has been linked to illegal habitat destruction and to the slaughter of great apes, a fact that has spurred 46 zoos to host cell phone recycling programs.[17]

But recycling computers and cell phones is not necessarily a simple solution. The Basel Action Network points out that e-waste is the fastest-growing waste stream in the industrial world, with much of it classified as hazardous waste.[18] Up to 80 percent of the electronics turned in for recycling in the United States is actually exported to developing countries, violating an international agreement designed to prevent irresponsible trade in hazardous waste and exposing workers in countries such as Nigeria and China to horrific environmental conditions at growing e-waste dumps.[19] In a positive step, several dozen e-waste recyclers have signed the Electronic Recycler's Pledge of True Stewardship, stating that they refrain from shipping e-waste abroad or using prison labor in the recycling process.[20]

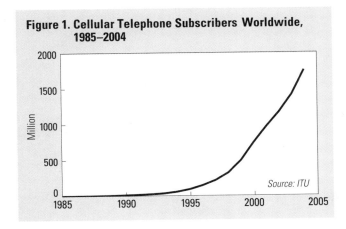

Figure 1. Cellular Telephone Subscribers Worldwide, 1985–2004

Source: ITU

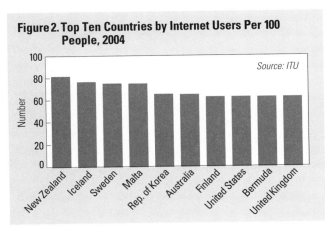

Figure 2. Top Ten Countries by Internet Users Per 100 People, 2004

Source: ITU

Cellular Telephone Subscribers and Internet Host Computers Worldwide, 1985–2005		
Year	Cellular Phone Subscribers	Internet Host Computers
	(million)	
1985	1	0.00
1986	1	0.01
1987	2	0.03
1988	4	0.08
1989	7	0.16
1990	11	0.38
1991	16	0.73
1992	23	1.31
1993	34	2.22
1994	56	5.85
1995	91	14.35
1996	145	21.82
1997	215	29.67
1998	319	43.23
1999	492	72.40
2000	741	109.57
2001	965	147.34
2002	1,168	171.64
2003	1,414	233.10
2004	1,757	317.65
2005 (prel)	n.a.	394.99

Source: ITU, Internet Systems Consortium.

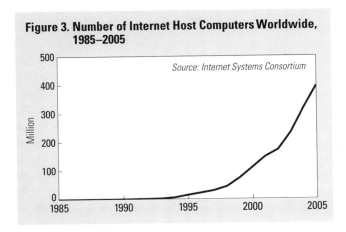

Figure 3. Number of Internet Host Computers Worldwide, 1985–2005

Source: Internet Systems Consortium

Health and Social Trends

J. Shadid, courtesy of USAID

Sewing a garland of HIV-awareness ribbons for use in an HIV-prevention program, Mali

▶ Population Continues to Grow

▶ HIV/AIDS Threatens Development

▶ Infant Mortality Rate Falls Again

For data and analysis on health and social trends, including refugees, cigarette production, and urbanization, go to www.worldwatch.org/vsonline.

Population Continues to Grow

Danielle Nierenberg

The world's population topped 6.4 billion in 2005.[1] (See Figure 1.) This is more than twice as many people as populated Earth in 1950.[2] Although the actual population growth rate continues to drop, from a high of over 2 percent in the 1970s to just above 1 percent today (see Figure 2), the number of people living on Earth increases by millions every year.[3] In 2005, world population grew by some 74 million people.[4] (See Figure 3.)

Ninety-five percent of population growth is in the developing world, and most of the world's people are added in just a few countries.[5] In fact, more than half of the world's people live in Asia.[6] China and India alone account for more than one third of total population growth, and 4 out of every 10 people on Earth are Chinese or Indian.[7] Today, there are more than 1.3 billion people living in China, while India's population is just over 1 billion.[8] By 2050, India is expected to surpass China in population and will account for more than 30 percent of the people in the world.[9] And the populations of other Asian countries, including Pakistan and Bangladesh, are expected to double by 2050.[10]

LINKS pp. 76, 78, 114

Cities in China and India are growing particularly fast. Although nearly two thirds of Chinese and Indians live in rural areas, both countries are experiencing some of the fastest rural-to-urban migration in history.[11] According to the United Nations Population Division, by 2007 more people will be living in cities around the world than in rural areas.[12] India already has more than 30 cities with populations over 1 million.[13]

Population is projected to at least triple by 2050 in Afghanistan, Burkina Faso, Burundi, Chad, Congo, the Democratic Republic of Congo, the Democratic Republic of Timor-Leste, Guinea-Bissau, Liberia, Mali, Niger, and Uganda.[14] These countries are among the least developed nations on the planet; the prospect of populations three times the size they are today does not bode well for improving that situation.[15]

Nine countries, according to the United Nations, are expected to account for half of the world's projected population increase by 2050: India, Pakistan, Nigeria, the Democratic Republic of Congo, Bangladesh, Uganda, the United States, Ethiopia, and China (listed in the order of the size of increase in national numbers).[16]

Despite Asia's high population, Africa continues to have the highest growth rate, increasing by 2.2 percent every year.[17] By 2050, Africa will have more than twice the number of people it has today, at over 1.9 billion.[18] While the fertility rate in the region has decreased overall in recent years, there are still some nations where the average woman has more than six children during her lifetime.[19]

On the other hand, in many of the world's industrial countries there is concern about low fertility rates. In some countries, including Canada, Italy, Japan, and Russia, populations are declining.[20] In Southern and Eastern Europe, fertility levels are at unprecedented lows of just 1.3 children per woman.[21] To encourage people to have more children, France offers tax incentives to couples willing to start a family.[22]

The nations with low fertility rates may be worrying about the "graying" of their populations as the average age of residents rises, but there are still more young people on Earth than ever before. Youth bulges—where people aged 15 to 29 account for nearly half of all adults—are found in 100 nations in the developing world.[23]

Lack of reproductive health care continues to prevent many people from planning and spacing births. Millions of couples still lack access to contraceptives and other family planning services.[24] At the same time, 500,000 women die each year due to complications from pregnancy and childbirth.[25] Getting maternal health care and family planning services to the women who need them can both save lives and ease the pressure of rapid population growth in some of the poorest nations of the world.

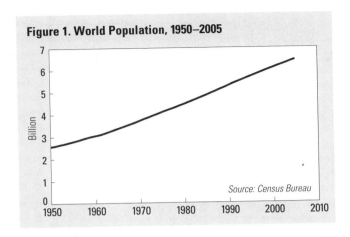

Figure 1. World Population, 1950–2005

Source: Census Bureau

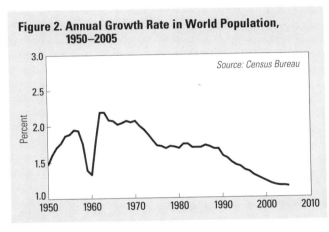

Figure 2. Annual Growth Rate in World Population, 1950–2005

Source: Census Bureau

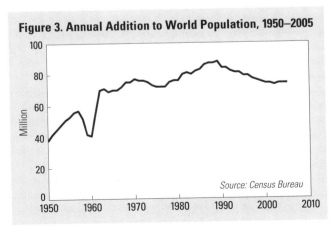

Figure 3. Annual Addition to World Population, 1950–2005

Source: Census Bureau

World Population, Total and Annual Addition, 1950–2005

Year	Total (billion)	Annual Addition (million)
1950	2.56	38
1955	2.78	53
1960	3.04	41
1965	3.35	70
1970	3.71	77
1971	3.79	77
1972	3.86	76
1973	3.94	75
1974	4.01	73
1975	4.09	72
1976	4.16	72
1977	4.23	72
1978	4.30	75
1979	4.38	76
1980	4.45	76
1981	4.53	80
1982	4.61	81
1983	4.69	80
1984	4.77	82
1985	4.85	83
1986	4.94	86
1987	5.02	87
1988	5.11	87
1989	5.19	88
1990	5.28	84
1991	5.37	84
1992	5.45	82
1993	5.53	81
1994	5.61	81
1995	5.69	79
1996	5.77	79
1997	5.85	77
1998	5.93	76
1999	6.00	75
2000	6.08	74
2001	6.15	74
2002	6.23	73
2003	6.30	74
2004	6.38	74
2005 (prel)	6.45	74

Source: U.S. Bureau of the Census.

HIV/AIDS Threatens Development

Lauren Sorkin

In 2005, approximately 5 million new HIV infections occurred, bringing the total number of people currently living with the virus to an estimated 40.3 million.[1] (See Figure 1.) About 3 million people died of AIDS-related illnesses in 2005; of these, nearly 600,000 were children under the age of 15.[2] Since the discovery of AIDS in 1981, the disease has claimed more than 37.1 million lives.[3] (See Figure 2.)

Sub-Saharan Africa continues to be the region most affected by the pandemic, accounting for 64 percent of new infections (more than 3 million people) in the last year.[4] (See Figure 3.) The Caribbean is the second most affected region; AIDS is the leading cause of death among people aged 15–44.[5] But the steepest increases in HIV infections in 2005 were recorded in Eastern Europe and Central Asia—with a 25-percent increase to 1.6 million infections in the former, primarily due to widespread intravenous drug use.[6]

LINKS pp. 74, 110

In Cambodia, just under 2 percent of the population is infected, including nearly 21 percent of sex workers, giving this country the highest HIV rate in Asia.[7] In China, Malaysia, and Viet Nam, the virus has spread rapidly; injected drug use is again the main route of transmission.[8] The World Health Organization estimates that without major efforts to raise awareness and lessen the stigma associated with AIDS, some 10 million Chinese may be infected in the next five years.[9]

Roughly 95 percent of people living with HIV/AIDS are of working age, making the disease a major developmental hurdle in highly affected countries.[10] The pandemic is especially a major threat to the world of work in Africa, where Kenya, Tanzania, Ethiopia, Mozambique, and Zimbabwe are home to 70 percent of the world's HIV-positive workforce.[11] According to the International Labour Organization, by 2005 the global workforce had lost 28 million workers to AIDS since the epidemic started, resulting in gross domestic product decreases worth $25 billion a year in the most affected countries in Africa, Asia, and the Caribbean.[12]

Economic losses tell only a small part of a story that includes erosion of social networks and human capital in developing countries affected by HIV. Conflict and corruption hasten the spread of HIV/AIDS because of a lack of safe blood supply, a shortage of clean needles for intravenous drug users, an insufficient supply of condoms and health care, and an increase in the number of displaced people—especially women and children—and in their vulnerability to sexual abuse and violence.[13]

Many of the 12 million AIDS orphans in sub-Saharan Africa are forced into child labor in order to support their families and must assume roles left vacant by parents prematurely taken by the virus.[14] The number of orphans is overwhelming traditional intergenerational support systems and leading to food scarcity, especially in rural areas.[15]

Although new initiatives hold promise, poor surveillance due to lack of public awareness and heavy social stigmas continues to be a major impediment in prevention and treatment for groups most vulnerable to HIV/AIDS, including men who have sex with men, sex workers, and injecting drug users.[16] Even in industrial countries, although HIV testing is becoming more routine, 27 percent of HIV-infected people in the United States and up to one third in the United Kingdom remain undiagnosed.[17]

For those diagnosed with HIV, access to care is still severely lacking, especially in developing countries. According to the World Health Organization, the number of individuals qualified for antiretroviral therapy has tripled to about 1 million since 2003—but nearly 90 percent of those eligible for treatment still do not have access to it.[18]

International funding available to fight the AIDS epidemic has risen from some $300 million in 1996 to nearly $8 billion in 2005.[19] Yet funding for treatment, testing, and counseling continues to fall short of the growing need. The Joint United Nations Programme on HIV/AIDS estimates that unless more funding is immediately dedicated to fighting the virus, a gap of $18 billion will emerge by 2007—severely limiting efforts to make antiretroviral drugs more available to those who need them.[20]

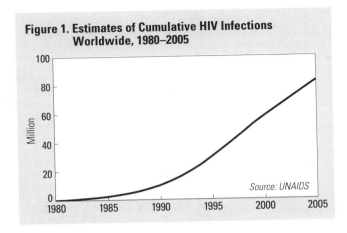

Figure 1. Estimates of Cumulative HIV Infections Worldwide, 1980–2005

Source: UNAIDS

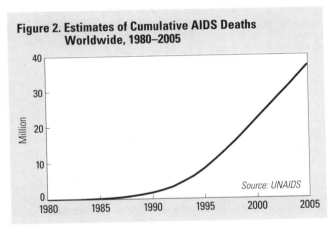

Figure 2. Estimates of Cumulative AIDS Deaths Worldwide, 1980–2005

Source: UNAIDS

Cumulative HIV Infections and AIDS Deaths Worldwide, 1980–2005

Year	HIV Infections	AIDS Deaths
	(million)	
1980	0.1	0.0
1981	0.3	0.0
1982	0.7	0.0
1983	1.2	0.0
1984	1.7	0.1
1985	2.4	0.2
1986	3.4	0.3
1987	4.5	0.5
1988	5.9	0.8
1989	7.8	1.2
1990	10.0	1.7
1991	12.8	2.4
1992	16.1	3.3
1993	20.1	4.7
1994	24.5	6.2
1995	29.8	8.2
1996	35.3	10.6
1997	40.9	13.2
1998	46.6	15.9
1999	52.6	18.8
2000	57.9	21.8
2001	62.9	24.8
2002	67.9	27.9
2003	72.9	30.9
2004	77.8	34.0
2005 (prel)	82.7	37.1

Source: UNAIDS.

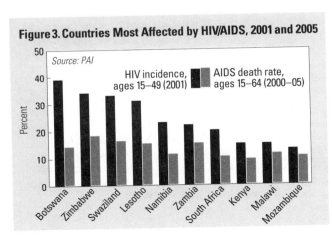

Figure 3. Countries Most Affected by HIV/AIDS, 2001 and 2005

Source: PAI

HIV incidence, ages 15–49 (2001)

AIDS death rate, ages 15–64 (2000–05)

Infant Mortality Rate Falls Again

Lauren Sorkin

The infant mortality rate—the number of children who die before one year of age per 1,000 live births—fell 7 percent in the last five years, from 61.5 deaths in 1995–2000 to 57.0 in 2000–05, its lowest level in history.[1] (See Figure 1.) Overall, however, the decline in infant mortality has slowed significantly recently—from

LINKS pp. 74, 118

an average drop of over 2 percent a year between 1950 and 1990 to less than 1 percent annually in the last 15 years, largely as a result of stagnation in improvements in health care and increases in infant mortality rates in some developing countries.[2]

In 2005, more than 4 million children died during their first year of life.[3] An additional 3.3 million children were stillborn, bringing the total number of infant deaths to more than 7 million a year.[4] Infant mortality rates for boys are higher than for girls worldwide, with 52 male children dying for every 1,000 live births in 2005 compared with 48 females.[5] Numerous factors contribute to the higher male infant mortality rate, including the fact that males have weaker immune systems and infant respiratory systems as well as a higher risk for premature birth, but more research is needed on this.[6]

Disparities in infant deaths between industrial and developing countries are increasing a bit (see Figure 2), largely as a result of the HIV/AIDS epidemic.[7] The countries most affected by the virus have seen a doubling of infant mortality rates in recent years.[8] The highest regional rate in 2005—94.2 deaths per 1,000 live births—was found in Africa.[9] (See Figure 3.) In addition to the devastation from HIV/AIDS in the region, it is estimated that 5,500 children die each day in Eastern and Southern Africa largely from preventable causes such as dehydration, malaria, and diarrhea.[10]

Other causes of infant mortality include diseases such as measles, tetanus, and tuberculosis that could be prevented if more children had access to basic immunization services.[11] While immunization initiatives have increased globally among children under 1, coverage varies among the regions of the world, with only 55 percent of children in Africa and 54 percent in

Asia getting immunized for measles.[12]

Access to prenatal and early infant care continues to be a major barrier to reducing infant mortality further in some of the world's least stable regions. In the 75 countries that account for most infant and child morality (70 percent of which are in Africa and Southeast Asia), only 43 percent of mothers receive any maternal or neonatal care.[13] According to the World Health Organization, increasing health care access in these countries in order to meet the Millennium Development Goals would require an investment of $52 billion, boosting public expenditures on health care from 6 to 18 percent per year.[14] An undersupply of adequately trained health care professionals, an inability to pay for health care services, the physical distance to health care facilities, and inadequate information about the importance of prenatal care are all significant hurdles to providing adequate mother and child care.[15]

Disparities between racial and ethnic groups, regional populations, and groups with differing levels of education are also visible around the globe. In Russia, for instance, children born to Kazakh parents are 1.5 times more likely to die than those born to Russian parents.[16] In the United States, the infant mortality rate was 6.4 per 1,000 live births in 2004 but 13.8 among African Americans, and it was far higher in the South than in the North.[17] Marginalization of ethnic groups contributes to higher rates of infant mortality, especially in countries like Ethiopia, Bangladesh, and Afghanistan that suffer from civil unrest, where conflicts contribute to people's exclusion from health care and the breakdown of the health care system.[18]

These disparities among different groups and areas in the world and the slowdown in reductions in the infant mortality rate during the 1990s led governments in 2000 to adopt as one of eight Millennium Development Goals the objective of reducing by two thirds, between 1990 and 2015, the under-five mortality rate.[19] Unfortunately, there has been little progress to date on reaching this goal.

Figure 1. World Infant Mortality Rates, 1950–2005

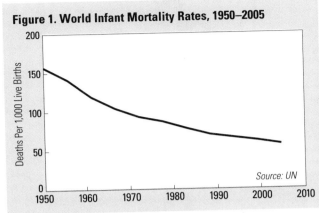

World Infant Mortality Rates, 1950–2005

Time Period	Rate
	(deaths per 1,000 live births)
1950–55	157
1955–60	141
1960–65	119
1965–70	104
1970–75	93
1975–80	87
1980–85	78
1985–90	70
1990–95	66
1995–2000	62
2000–05	57

Source: UN.

Figure 2. Infant Mortality Rates in Industrial and Developing Countries, 1950–2005

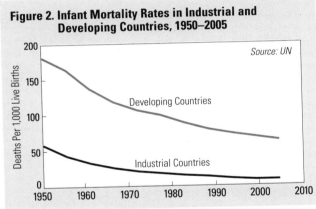

Figure 3. Infant Mortality Rate by Region, 2000–05

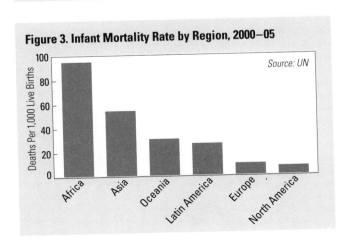

Conflict and Peace Trends

© Claire McEvoy/IRIN

Refugee women line up for food in Darfur, Sudan

▶ Number of Violent Conflicts Drops

▶ Military Expenditures Keep Growing

▶ Peacekeeping Expenditures Set New Record

For data and analysis on conflict and peace trends, including nuclear arsenals and resource conflicts, go to www.worldwatch.org/vsonline.

Number of Violent Conflicts Drops

Michael Renner

During 2005, the number of wars and armed conflicts worldwide declined to 39—the lowest since peaking in the early 1990s, according to AKUF, a conflict research group at the University of Hamburg.[1] (See Figure 1.) The number of full-fledged wars stood at 28, up slightly from 27 in 2004.[2] But the number of armed conflicts declined from 15 to 11; these are of lesser severity or for other reasons do not meet the criteria to be considered full-scale wars.[3]

Two new conflicts erupted in 2005, in the Democratic Republic (DR) of the Congo and in Saudi Arabia, but the guns fell silent in five old conflict locations. These included Chad, Georgia's Abkhazia and South Ossetia regions, and the Ituri district and Kivu provinces of the DR Congo.[4] The issues driving these conflicts remained unresolved, however. Although hostilities among various militias in eastern Congo died down, they were replaced by fresh fighting that was triggered by the central government's efforts to reassert control.[5]

LINKS p. 86

Another conflict still listed in AKUF's 2005 report—in Indonesia's Aceh province—was ended by a peace agreement signed in August 2005 between the central government and the Free Aceh Movement.[6] The devastating tsunami of December 2004 convinced those in conflict that peace was essential for reconstruction to take place.[7] The first phase of the agreement was carried out successfully. On 21 December 2005, the last rebel weapons to be decommissioned were cut apart in a public ceremony, and the last government forces to be withdrawn pulled out soon thereafter.[8]

The trend in armed conflicts worldwide has been on a downward trajectory since the early 1990s. This represents a stunning turnaround from the inexorably rising numbers of earlier decades. Still, given incomplete and contradictory information, it remains difficult to tally the number of conflicts—and particularly the number of victims—with precision. And the definitions and methodological tools developed by conflict researchers can at times inhibit a full accounting. For instance, fighting that does not involve government forces on at least one side is excluded from most data sets.

The Uppsala/PRIO Armed Conflict Dataset is one of the most respected sources of information.[9] (See Figure 2.) Yet for reasons of methodology it fails to capture certain conflicts (such as the fighting in Nigeria's Niger Delta in recent years), and it records some—such as the Rwandan genocide of 1994 or mass violence in the DR Congo after 2001—at much lower levels of intensity than appears warranted.

Too little is known about how many people fall victim to organized violence. Most data collection efforts focus on battle deaths, while civilian deaths—many the result of failing economies and collapsing health systems—are typically neglected. Information on the number of civilians killed is often patchy or contradictory. It is particularly difficult to find out how many people perish through the indirect impacts of warfare.

The DR Congo may have suffered the highest death toll of any recent war. A study in *The Lancet* estimates that 3.9 million people have died there since 1998 and that almost 38,000 civilians continue to die every month.[10] Less than 2 percent of the deaths were the direct result of violence; most people succumbed to disease and hunger.[11]

Undertaking a broad assessment, researchers at the Heidelberg Institute for International Conflict Research in Germany counted 249 political conflicts worldwide in 2005.[12] Of these, 24 involved a high level of armed violence and 74 some occasional violence. The remainder—61 percent—were carried out without resort to violence.[13]

Conflicts sometimes end because one side scores a decisive victory; at other times the fighting simply peters out. And in still other cases there are dedicated efforts to find a negotiated solution, including peacekeeping efforts and international sanctions. In 2005, informal talks or formal negotiations took place in 24 conflicts and led to 22 ceasefire agreements or peace treaties in places like Burundi, Senegal, Sudan, and Aceh.[14] U.N. sanctions remained in place against nine states: Afghanistan, Côte d'Ivoire, DR Congo, Iraq, Liberia, Rwanda, Sierra Leone, Somalia, and Sudan.[15]

Figure 1. Wars and Armed Conflicts, 1950–2005

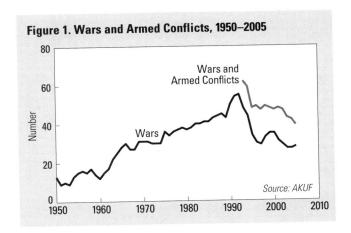

Source: AKUF

Figure 2. Armed Conflicts, 1950–2004

Source: Gleditsch et al.

Wars and Armed Conflicts, 1950–2005

Year	Wars	Wars and Armed Conflicts
		(number)
1950	13	
1955	15	
1960	12	
1965	28	
1970	31	
1975	36	
1976	34	
1977	36	
1978	37	
1979	38	
1980	37	
1981	38	
1982	40	
1983	40	
1984	41	
1985	41	
1986	43	
1987	44	
1988	45	
1989	43	
1990	50	
1991	54	
1992	55	
1993	48	62
1994	44	·59
1995	34	48
1996	30	49
1997	29	47
1998	33	49
1999	35	48
2000	35	47
2001	31	48
2002	29	47
2003	27	43
2004	27	42
2005 (prel)	28	39

Source: AKUF and Institute for Political Science, University of Hamburg.

Military Expenditures Keep Growing
Michael Renner

World military expenditures hit $1,024 billion in 2004, the most recent year for which data are available.[1] (See Figure 1.) After the end of the cold war, these expenditures initially declined. But after 1998, and especially after the September 2001 attacks in the United States, they expanded once again—by an annual average of 6 percent since 2002.[2]

The steepest increases, in percentage terms, occurred in Central Asia, North Africa, South Asia, and North America. Only in Central and South America and in Western Europe did expenditures rise less than 10 percent.[3]

As the Stockholm International Peace Research Institute points out, real military spending is likely to be far higher than it reports, as its numbers reflect official government data on budgeted expenditures, but actual outlays are sometimes higher. And many national armed forces rely on additional off-budget income, often derived from military-owned businesses.[4]

LINKS pp. 82, 86

While the United States, Russia, China, India, Israel, and Saudi Arabia were prominent among countries revving up their spending, Brazil, Mexico, and Germany were notably among those cutting back.[5] The United States spends almost as much as the rest of the world combined: $478 billion in 2004, or 47 percent of the global total.[6] Included in this figure are the costs of U.S. military operations in Afghanistan and Iraq and the "war on terror," which have been funded through supplementary appropriations, not the regular military budget. In fiscal years 2001–05, these funds added up to $347 billion.[7] Another $72.4 billion has been requested for fiscal year 2006.[8]

Even though U.S. military spending is at its highest level since 1969, the U.S. armed forces (soldiers and civilians) stood at only 2.1 million people in 2005, less than half the number of 40 years ago. Employment in the arms industry, however, stands at 3.9 million, compared with 2.9 million in 1969.[9]

The United Kingdom, France, Japan, and China account for $180 billion, or 18 percent of global expenditures. They are followed by Germany, Italy, Russia, Saudi Arabia, and South Korea ($122 billion total, 12 percent), then India, Israel, Canada, Turkey, and Australia ($59 billion, 6 percent). The top 15 budgets together run to $839 billion, or 82 percent of the global total.[10] (See Figure 2.) NATO countries are the dominant spenders.

The budgets of Russia, China, and India remain difficult to assess accurately. Apart from a lack in transparency, there is considerable difficulty in properly measuring the resources that these countries devote to their military sectors. Expressing their budgets in dollars using market exchange rates tends to understate their capacity to maintain armed forces. An alternative measure, so-called purchasing power parity rates (which are best suited for measuring domestic purchases), would on the other hand considerably overstate the ability of these countries to shop for weapons, and hence their military prowess.[11]

With rising military budgets, the arms business is booming: in 2003, the top 100 weapons-producing companies (excluding China) sold arms worth $248 billion, up from $211 billion the year before.[12] U.S. companies accounted for $157 billion in sales, and European firms for $76 billion.[13]

In a world where billions of people struggle to survive on $1–2 per day, governments spend on average $166 per person on weapons and soldiers.[14] But this average conceals tremendous regional differences. In North America, per capita expenditures in 2004 were $1,487; in Western Europe, $542; in the Middle East, $254; in Asia, $46; and in Africa, a mere $18.[15]

In 2004, industrial countries in the Organisation for Economic Co-operation and Development spent $830 billion on military programs—more than 10 times what they spent on development assistance ($81.4 billion).[16] (See Figure 3.) For the United States, the military spending–foreign aid ratio was a whopping 24:1.[17]

At a time when endemic poverty, health epidemics, climate change, and mass unemployment cry out for attention, the continued growth in military budgets reflects a troubling set of priorities and a failure to address the underlying reasons for much of the world's instability.

Figure 1. World Military Expenditures, 1950–2004

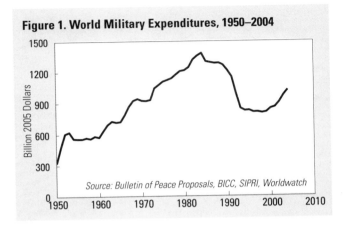

Source: Bulletin of Peace Proposals, BICC, SIPRI, Worldwatch

Figure 2. Military Spending, Selected Countries and Areas, 2004

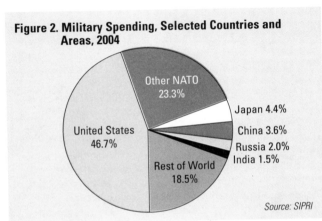

Source: SIPRI

Figure 3. Ratio of Military Spending to Official Development Assistance, Selected Coutries, 2004

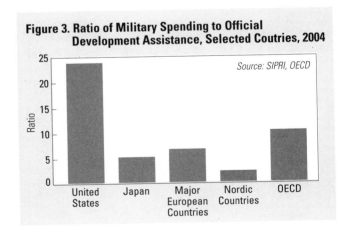

Source: SIPRI, OECD

World Military Expenditures, 1950–2004

Year	Expenditure (billion 2005 dollars)
1950	323
1955	558
1960	574
1965	723
1970	925
1971	922
1972	932
1973	1,045
1974	1,077
1975	1,110
1976	1,125
1977	1,142
1978	1,177
1979	1,210
1980	1,218
1981	1,247
1982	1,321
1983	1,357
1984	1,383
1985	1,303
1986	1,294
1987	1,286
1988	1,288
1989	1,269
1990	1,217
1991	1,154
1992	989
1993	848
1994	826
1995	829
1996	811
1997	813
1998	804
1999	812
2000	847
2001	860
2002	908
2003	974
2004	1,024

Source: Worldwatch, Bulletin of Peace Proposals, BICC, SIPRI.

Peacekeeping Expenditures Set New Record   *Michael Renner*

The budget for United Nations peacekeeping operations from July 2005 to June 2006 is a record $5 billion—climbing past the previous peak of $4.6 billion in 2004–05.[1] (See Figure 1.) Some 70,000 soldiers, military observers, and police served in 16 peacekeeping missions at the end of 2005.[2] (See Figure 2.) Including international and local civilian staff and volunteers, total personnel came to about 85,000.[3] (See Figure 3.)

The United Nations also maintains 10 smaller "political and peace-building" missions, with a mostly civilian staff of 2,349 as of late 2005.[4] The largest of these are in Afghanistan (set up in March 2002), Iraq (August 2003), and Timor-Leste (May 2005).[5]

p. 82

Pakistan, Bangladesh, India, Nepal, and Sri Lanka together accounted for a stunning 42 percent of all peacekeepers in 2005.[6] Six sub-Saharan African countries—Ethiopia, Ghana, Nigeria, South Africa, Senegal, and Kenya—together provided another 22 percent.[7] Jordan, Uruguay, Morocco, and Brazil rounded out the top tier of contributors with another 13 percent.[8]

Western countries now account for less than 10 percent of all peacekeeping personnel, down from 45 percent in 1998.[9] The permanent members of the Security Council—China, France, Russia, the United Kingdom, and the United States—are in a powerful position to influence the missions, but they collectively provide a mere 4 percent of personnel.[10] In fact, Rwanda now provides more peacekeepers than Russia does.[11]

The mission in Sierra Leone, once one of the largest, with 17,500 peacekeepers, ended in December 2005 after six years.[12] Deployments in the Democratic Republic of the Congo (with more than 19,000 military, police, and civilian staff) and in Liberia (about 17,800 personnel) are now by far the largest missions.[13] They are followed by missions in Haiti (about 10,000 staff) and Côte d'Ivoire (about 8,500).[14] Operations in Kosovo, Burundi, and Sudan each have between 5,500 and 6,500 staff.[15] The Sudan mission is to expand to close to 15,000.[16]

U.N. Secretary-General Kofi Annan has asked that 3,875 military and police be added to the overstretched mission in Côte d'Ivoire.[17]

But the United States and Japan were against earlier proposals for an expansion, for financial reasons.[18] Top U.N. envoy Jan Pronk has called for a U.N. peacekeeping force of 12,000–20,000 people in Sudan's Darfur region to replace an African Union force that is faltering in the face of growing violence.[19]

Two operations explain the recent upsurge in peacekeeping budgets. The U.N. General Assembly approved $1.15 billion for the Congo mission for 2005–06 and $970 million for Sudan.[20] Yet peacekeeping finances continue to be problematic as member states pay their dues late or not in full. By December 2005, arrears stood at about $2.9 billion, above the December 2004 figure of $2.57 billion.[21] Japan was the single largest debtor, with $845 million in arrears, followed by the United States at $843 million.[22] South Korea, Italy, Germany, France, Spain, the United Kingdom, and China are the next largest debtors, together owing $672 million.[23]

As opposed to earlier stances of strict neutrality even in the face of mass killings (as in Rwanda and Bosnia), some recent missions have embraced far more aggressive tactics, particularly in Haiti and the Congo.[24] This shift has taken place against the backdrop of ongoing discussions about the international community's "responsibility to protect" civilians in war zones.[25] One consequence has been higher fatalities among peacekeepers. Some 353 peacekeepers died in 2003–05, the highest number since 1995.[26]

Non-U.N. peacekeeping and monitoring missions can also be found in all regions of the world. During 2005, more than 40 missions were supported by regional organizations and by ad hoc coalitions of countries.[27] Altogether, they involved an estimated 51,000 soldiers.[28] The largest of these continue to be deployments in Kosovo, Afghanistan, and Bosnia led by NATO and the European Union.[29] But sometimes even small missions can be highly effective. For example, the Aceh Monitoring Mission, with just 230 staff, oversees a peace agreement ending a 29-year conflict in Indonesia's Aceh province.[30]

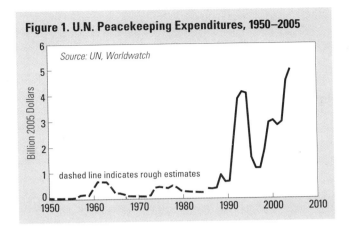

Figure 1. U.N. Peacekeeping Expenditures, 1950–2005

Source: UN, Worldwatch

dashed line indicates rough estimates

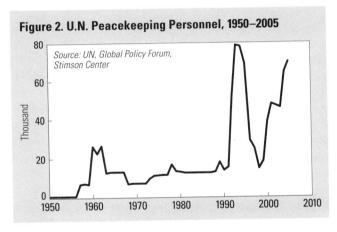

Figure 2. U.N. Peacekeeping Personnel, 1950–2005

Source: UN, Global Policy Forum, Stimson Center

U.N. Peacekeeping Expenditures, 1986–2005

Year	Expenditure
	(billion 2005 dollars)
1986	0.378
1987	0.365
1988	0.391
1989	0.899
1990	0.632
1991	0.646
1992	2.276
1993	3.850
1994	4.119
1995*	4.063
1996*	1.577
1997*	1.154
1998*	1.153
1999*	1.872
2000*	2.926
2001*	3.009
2002*	2.812
2003*	2.951
2004*	4.558
2005* (prel)	5.003

* July to June of following year.

Source: U.N. Department of Public Information and Worldwatch Institute database.

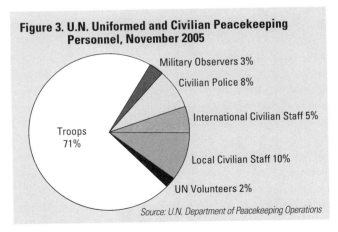

Figure 3. U.N. Uniformed and Civilian Peacekeeping Personnel, November 2005

Military Observers 3%

Civilian Police 8%

International Civilian Staff 5%

Troops 71%

Local Civilian Staff 10%

UN Volunteers 2%

Source: U.N. Department of Peacekeeping Operations

Part Two

SPECIAL FEATURES

Environment Features

Mangrove Action Project

Fishing beside mangroves, Sulawesi, Indonesia

▶ Global Ecosystems Under More Stress

▶ Coral Reef Losses Increasing

▶ Birds Remain Threatened

▶ Plant Diversity Endangered

▶ Disappearing Mangroves Leave Coasts at Risk

▶ Deforestation Continues

▶ Groundwater Overdraft Problem Persists

▶ Reducing Mercury Pollution

For data and analysis on these environmental topics and on air pollution, sea level rise, and global ice melt, go to www.worldwatch.org/vsonline.

Global Ecosystems Under More Stress

Erik Assadourian

In 2005, the Millennium Ecosystem Assessment (MA)—a comprehensive analysis produced by 1,360 scientists after four years of consultations and research—determined that the health of the world's ecosystems was in significant decline.[1] Ecosystems provide essential services to people. Yet of the 24 ecosystem services examined in the MA—including provision of fresh water, food, and fiber and the regulation of climate and air quality—the scientists found that 15 (62.5 percent) are being degraded or used unsustainably, a trend that "could grow significantly worse during the first half of this century."[2] While the report noted that overexploitation of many of these services led to net gains in economic development, it also made clear that if the degradation is left unaddressed, the ability of ecosystems to provide these benefits in the future will be diminished.[3]

LINKS pp. 94, 96, 98, 100, 102

Degradation of many of Earth's natural systems has been brought about by human activity, according to the report. As the MA outlines, approximately 40 percent of the world's coral reefs have been lost or degraded, water withdrawals from rivers and lakes have doubled since 1960, the atmospheric concentration of carbon dioxide has jumped 19 percent since 1959, and the global species extinction rate has increased as much as 1,000 times over the typical rate seen across Earth's history.[4]

Moreover, decline of these systems is also increasing the risk of "nonlinear change"—abrupt, disruptive, and potentially irreversible changes such as regional climate shifts, the collapse of fishery resources, the emergence of new diseases, and the formation of dead zones in coastal waters.[5] The weakening of these systems is also exacerbating poverty among some groups of people—a trend that could worsen dramatically if abrupt changes are unleashed.[6]

Indeed, the conclusions of the report were so dire that the MA Board of Directors noted in its own summary statement that "human activity is putting such strain on the natural functions of Earth that the ability of the planet's ecosystems to sustain future generations can no longer be taken for granted."[7]

Other indicators confirm that humans are exceeding the capacity of Earth's systems. According to the Living Planet Index—which measures the state of the world's biodiversity by tracking population trends for more than 1,100 species—biodiversity declined 40 percent between 1970 and 2000.[8]

The Ecological Wellbeing Index, an average of 51 environmental indicators, found that few countries are ecologically healthy.[9] None of the 180 countries looked at received a "good" rating; only 27 scored a "fair," while 81 were rated "medium," 68 "poor," and 4 "bad."[10]

The Ecological Footprint, a conservative measure of natural resource consumption, also shows that humans are putting considerable pressure on Earth.[11] This indicator calculates the total amount of land the world's countries need to produce the resources they use (including food and fiber), to absorb the waste they generate from energy used, and to provide space for infrastructure.[12] In 2002, humanity used the equivalent of 13.7 billion "global hectares" of biocapacity—2.5 billion more than Earth's biologically productive area of 11.2 billion global hectares.[13] This translates into humanity overdrawing the natural capital it depends on by 23 percent in 2002.[14]

The Ecological Footprint shows that humanity has been living beyond its means since 1987 and thus drawing down the ecological capital that is the basis for the continued health of the planet.[15] (See Figure 1.)

Of course, some countries are using far more biocapacity than others. If all humans were to consume at high-income-country levels, another 2.5 planets would be needed.[16] Said another way, with everyone consuming at this level, Earth could sustain only 1.8 billion people—not today's population of 6.5 billion.[17] (See Table 1.) And worse, at U.S. consumption levels the planet could support just 1.2 billion people.[18]

With world population projected to hit 9.2 billion in 2050, with rapid economic growth in major developing economies such as China and India, and with continued high consumption rates in industrial countries, ecosystem degradation is likely to accelerate in coming decades.[19]

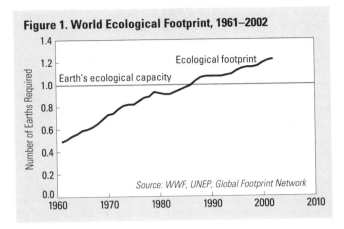

Figure 1. World Ecological Footprint, 1961–2002

Source: WWF, UNEP, Global Footprint Network

Yet another measure, the 2006 Environmental Performance Index, ranked the performance of the world's governments in achieving 16 crucial environmental goals.[22] The indicators chosen included those that play a key role in environmental health and ecosystem vitality, such as air pollution, water consumption, agricultural subsidies, energy efficiency, and wilderness protection.[23]

According to the Environmental Performance Index, only 6 of the 133 countries that researchers measured achieved 85 percent or higher in 2005 for their efforts, with New Zealand, Sweden, and Finland leading the way.[24] At the bottom of the list, 24 countries scored less than 50 percent; 16 of these countries were in sub-Saharan Africa, in large part because of health and sanitation deficiencies.[25]

Perhaps most useful is the Environmental Performance Index's categorization of countries by similarity, such as low-income, primarily desert, or high-population density.[26] Comparing similar countries shows that certain constraints do not have to prevent environmental success. For example, Japan—a high-population-density country—still scored 82 percent, suggesting, the authors note, that "demography is not destiny."[27] With the right policies, low-performing countries, whether impoverished, primarily desert, or densely populated, could improve environmental health and sustainability records if they make these their priorities.

But with aggressive policy responses, such as the elimination of harmful subsidies, significant investments in environmentally responsible technologies, and the creation of taxes on activities that have externalized ecological costs, it is possible to reverse some of the decline.[20] The world, however, has not come up with the political commitment to make the needed changes. According to the 2006 Global Governance Initiative report, the world's governments, businesses, and civil society received a score of 2 out of 10 in dealing with environmental issues, as there was only minimal success in addressing climate change, halting biodiversity decline, and providing clean water and sanitation.[21]

Table 1. Sustainable Population at Different Consumption Levels

Consumption Level	Biocapacity Used Per Person	Sustainable Population at this Level
	(global hectares)	(billion)
High income	6.4	1.8
Middle income	1.9	5.9
Low income	0.8	14.0
Global Average	2.2	5.1

Source: Global Footprint Network.

Coral Reef Losses Increasing

Lisa Mastny

As of late 2005, an estimated 20 percent of the world's coral reefs had been "effectively destroyed," showing live coral losses of at least 90 percent and no immediate prospects for recovery, according to the Global Coral Reef Monitoring Network.[1] Another 24 percent face imminent risk of collapse as a result of human pressures and 26 percent face longer-term loss—bringing the share of world reefs now threatened or destroyed to 70 percent, up from 59 percent in 2000.[2] The greatest destruction has occurred in the Caribbean Sea, Indian Ocean, Gulf regions, and Southeast Asia.[3] (See Table 1.)

LINKS pp. 42, 92

An estimated half-billion people in more than 100 countries—many of them small islands—rely on coral reefs for essential goods and services.[4] During the 2004 Indian Ocean tsunamis, reefs provided a natural barrier against wave damage, erosion, and flooding.[5] Reefs are important feeding and breeding grounds, supplying commercial fisheries and serving as a primary food source for some 30 million people.[6] They also generate significant tourism revenue, with Florida's reefs estimated to bring in $1.6 billion annually.[7] The U.N. Environment Programme estimates the overall value of coral reefs at $100,000–600,000 per square kilometer a year.[8]

These unique assemblages of tiny coral animals and symbiotic plants cover less than 0.1 percent of ocean area but are among Earth's most complex and productive ecosystems.[9] They provide habitat for as many as 1 million species, including more than a quarter of known marine fish species.[10] Any number of antibiotics, cancer treatments, and HIV drugs have been developed from molecules derived from reefs.[11]

Most of the world's coral reefs have been lost as a result of direct human pressures—from fishing, mining of coral, coastal development, waste dumping, vessel collisions, oil spills, and runoff from inland deforestation and farming.[12] Human threats will only worsen: nearly 75 percent of coral reefs are found in developing regions with rapidly expanding populations.[13]

The booming demand for reef species for food and the global aquarium trade has depopulated many coral ecosystems.[14] A survey of over 1,100 reefs between 1997 and 2001 found that many high-value species were missing from shallow areas where they were once abundant, with spiny lobster gone from 83 percent of Atlantic reefs, Nassau grouper from 82 percent of Caribbean reefs, and Barramundi cod from 95 percent of Indo-Pacific reefs.[15] In many reefs in East Africa, South and Southeast Asia, and the Caribbean, it is now rare to see a fish more than 10 centimeters long.[16]

Common fishing methods are often highly destructive to reefs. In Southeast Asia, a 1-kilogram "bottle bomb" can leave a crater of rubble 1–2 meters in diameter, killing 50–80 percent of the coral in areas regularly hit.[17] These "blast fishing" practices have degraded an estimated 75 percent of Indonesia's reefs.[18] And in the Philippines, more than a million kilograms of cyanide have been injected into reefs to stun fish since the 1960s—a procedure that harms many nontarget species as well.[19] Powerful trawling vessels can grab up a quarter of seabed life in a single pass.[20]

The biggest threat to reefs today, however, is climate change, which can exacerbate the effects of other stresses.[21] Reefs are like the proverbial canaries in a coal mine: early indicators of life-threatening conditions.[22] Microscopic plants (zooxanthellae) live in the tissue of coral animals and provide them with food and color; if ocean temperatures rise even 1 degree Celsius above normal, these plants can come under stress.[23] If the problem continues, the corals expel the plants and turn chalky white, often dying.[24]

Such "bleaching" events have increased in frequency and intensity since the early 1980s.[25] In 1998, a combination of El Niño/La Niña–related climatic changes and record-high tropical sea surface temperatures caused the worst episode on record, destroying some 16 percent of the world's reefs, including thousand-year-old corals as deep as 40 meters.[26] Indian Ocean reefs suffered losses of up to 90 percent in some areas, with damages estimated at between $700 million and $8.2 billion.[27] Other localized bleaching events occurred in 2000 and 2003.[28]

Table 1. Status of Coral Reefs Around the World

Location	Share Destroyed (percent)	Reef Condition
Pacific Ocean (39 percent of total reef area)	2 in Australia and Papua New Guinea; 14 in rest	Generally good, with remote reefs off Polynesia among the least degraded worldwide. Southwest Pacific saw extensive bleaching in 2000–02, and reefs of New Caledonia, Samoa, Solomon Islands, and Vanuatu have suffered cyclone damage. Other threats include development, sediment and nutrient runoff, overfishing, and predation by crown-of-thorns starfish. Reef protection is strong in Fiji and Australia but lacking in Papua New Guinea.
Southeast and East Asia (34 percent)	38 in Southeast Asia; 14 in rest	Continues to decline, particularly in the Philippines, Viet Nam, Thailand, and Singapore. Main threats include overfishing and destructive fishing, land-based pollution, coastal development and sedimentation, deforestation, and dredging and reclamation of reef areas.
Indian Ocean (11 percent)	45 in South Asia; 34 in rest	Bleaching in 1998 caused widespread damage, particularly in the Maldives, Sri Lanka, and parts of western India, though recovery is encouraging. In East Africa, regrowth is estimated at 30–50 percent. The greatest near-term threat is from growing coastal populations, including land-based pollution and development, coral mining, and overfishing. Regional awareness of the need for management and conservation is growing.
Caribbean Sea and Atlantic Ocean (9 percent)	58	Major destruction, particularly of reef-building staghorn and elk corals. Live coral on many Caribbean reefs has declined by 80 percent in 25 years, with little sign of recovery. Threats include bleaching, hurricanes, overfishing, pollution, coastal modification, dredging, and mining. If degradation continues, annual economic losses in Caribbean could reach $350–870 million by 2015. Of 285 marine protected areas in the region, only 6 percent are considered effectively managed.
Middle East (8 percent)	65 in Gulf regions; 4 in Red Sea	Near-shore reefs in the Arabian Sea and Persian Gulf, destroyed by bleaching in 1996 and 1998, have shown little recovery. Red Sea reefs remain healthy but are threatened by shipping, tourism development, bleaching, and crown-of-thorns starfish.

Sources: World Conservation Monitoring Centre, Global Coral Reef Monitoring Network, and Environment News Service.

As of late 2004, approximately 40 percent of reefs seriously damaged in 1998 were either recovering soundly or had recovered, particularly those in well-managed or remote areas.[29] Unfortunately, such mass bleaching episodes are predicted to be more frequent over the next half-century.[30] A die-off may already be under way in the Caribbean, where unusually warm waters in 2005 led to bleaching of up to 95 percent of coral colonies in some areas.[31] Early evidence suggests that some of the world's reefs may be capable of adapting to such warming by forming new symbiotic relationships with more resilient algae.[32] But unless urgent protection measures are taken, half of all reefs could be lost in the next 40 years.[33]

One important solution is establishing marine reserves where fishing, anchoring, and other harmful activities are banned. In early 2004, Australia extended its protection of the Great Barrier Reef system from 5 percent to 33 percent, to cover 114,530 square kilometers.[34] And the United States aims to protect a fifth of its reefs in reserves by 2010.[35]

Birds Remain Threatened

Katie Carrus

BirdLife International estimates that 12 percent of all bird species—1,212 in all—fell in the "threatened" category in 2005.[1] And scientists predict that by 2100, 6–14 percent of all bird species will be extinct and 7–25 percent will be functionally extinct, a category that indicates only a handful of individual animals remain, most of whom are unable to reproduce due to health, age, or lack of mating partner.[2]

Birds perform critical ecosystem functions, and their continued disappearance spells disaster for the delicate web of life on which we all depend. In the Atlantic Forest of northeast Brazil, for example, nearly one third of the trees that require animal dispersal of seeds for reproduction are in peril because the large fruit-eating birds they depend on are threatened by habitat loss and overhunting.[3] Similarly, birds that help animals decompose are vital for human health. When the vulture population plummeted in India in 1997, the feral dog and rat populations exploded, providing the catalyst for 30,000 rabies deaths throughout the country.[4]

Habitat degradation and loss affect 86 percent of globally threatened birds.[5] And intensive agricultural practices are the main causes of habitat destruction. Massive conversion of land for agricultural use and high-input farming practices such as deep drainage, large-scale irrigation, heavy pesticide use, overgrazing, and monocropping are responsible for the degradation of agricultural and seminatural habitats and the decline of biodiversity around the world.[6] This industrial farming model is blamed for the decline of 38 percent of all bird species in Europe and threatens innumerable species as it spreads throughout the world.[7] (See Table 1.)

Avian habitat is also endangered by unsustainable forestry practices. Today, forest-dwelling parrots—a category that includes more than 80 percent of all parrot species—are on a drastic decline due to deforestation.[8] Similarly, mangrove forests in places like Borneo and Trinidad provide essential habitat for the scarlet ibis and straight-billed woodcreeper, yet both species are at risk because at least 35 percent of all mangrove forest area has been lost in the past two decades.[9]

The second major threat to birds is overexploitation, a problem that is connected to and reinforced by all the other threats to birds. Hundreds of thousands of birds are seized each year and sold in the global marketplace.[10] Parrots account for 117 of the world's globally threatened bird species.[11] Illegal trade in exotic parrots for the pet market nets millions of dollars around the world annually and is responsible for the death of four out of five parrots before they even arrive at their destination.[12] Thanks to this, one quarter of all parrot species are threatened—one of the highest proportions of any major species of birds.[13]

Overexploitation is not limited to the pet trade. BirdLife International reports that 262 bird species are threatened by hunting for food.[14] And the commercial fishing industry is responsible for an unsustainable "bycatch" beyond that of extra fish unintentionally caught up in nets: all 21 species of albatross are now threatened, an indication of how many seabirds get hooked on longline fishing lines and drown.[15]

The world's birds are also increasingly threatened by ecosystems that are imbalanced by the invasion of alien species. Closely tied to habitat loss because of the havoc foreign species wreak on ecological support systems, the problem of invasive species is widespread: nearly 30 percent of globally threatened bird species are affected by outsiders moving into their territory.[16]

Island species are particularly vulnerable to invasive species because of their isolated evolutionary history. Ninety percent of all birds that have become extinct since 1800 were island birds, and half of these were at least partially driven to extinction by invaders.[17] For example, only about 3,200 Christmas Island frigate birds in Australia—once the largest seabird breeding colony in the world—remain because of the introduction of an exotic ant species.[18] They are part of the 67 percent of all oceanic-island birds that are threatened.[19]

Pollution affects 12 percent of all globally threatened birds—187 species.[20] The impacts are both direct and indirect: pollution can reduce food supplies and degrade habitat; it can also kill birds outright or reduce their health or fertility.[21] Wetland habitats for birds, for instance, are frequently degraded by nutrient pollution from fish farms, sewage outfalls, and agricultural runoff (including livestock manure dumped directly into rivers). These can lead to wetlands and coastal habitats plagued with algal blooms, poisonous "red tides" that trigger fish kills, and permanently oxygen-less "dead zones," as has happened in parts of the Black Sea and the Gulf of Mexico.[22]

Pesticide and chemical residues from farms and gardens—even on private plots—disrupt ecosystems and reduce the number of insects and seeds available to birds like skylarks and song thrushes.[23] Landfill runoff, suspended sediment from excessive soil erosion (due to intensive agriculture), and the disproportionate presence of heavy metals, salts, and trace elements like selenium can all have a negative impact on birds like the white-tailed sea eagle in Sweden and the bald eagle in North America.[24]

A growing number of scientists believe that global climate change has affected birds directly through altered behavior, distribution, and patterns and timing of breeding and migration. Indirectly, climate change has affected habitats (altering growing seasons, for instance, or the distribution of species).[25] Temperature shifts over the past few decades have triggered marked delays in the arrivals and departures of migrating birds, for example, severely affecting their reproductive success.[26] And the U.S. Environmental Protection Agency reports that warmer summers are likely to alter the distribution of the bobolink, a North American songbird currently found throughout New England.[27] If atmospheric carbon dioxide levels double, as predicted, no bobolinks will be found south of the Great Lakes.

Finally, the physical landscape erected by people presents a host of challenges. More than

Table 1. Birds Threatened by Intensive Agricultural Practices, by Region

Region	Practice	Endemic Birds Threatened
South Africa	Overgrazing	Blue swallow
Southeast Asia	Excessive pesticide use	Malayan whistling-thrush
Europe	Large-scale irrigation	Red kite
North America	Monocropping; genetically modified organisms	Henslow's sparrow
South America	Forest clearing for agricultural exports	Bay-breasted cuckoo

Source: BirdLife International, Audubon Society.

1 billion birds strike windows every year in the United States—the birds' inability to see windows fatally affects at least 225 different species.[28] And an estimated 4–5 million birds are killed annually at 77,000 hazardous communication towers in the United States.[29]

Daunting as it may appear, the chance to improve prospects for the world's birds is real. Nongovernmental organizations—along with government agencies, scientific communities, and groups of concerned citizens—are working to reverse some of the most harmful trends and to promote healthy stewardship of the ecosystems and resources birds depend on. The rare Seychelles magpie robin's population has risen from 23 to about 154 birds since BirdLife International and the Royal Society for the Protection of Birds implemented a recovery program in the early 1990s. The plan included habitat creation, establishment of new populations, and the control of nonnative species.[30] If population numbers remain the same, the species could be downlisted from critically endangered status to endangered on the Red List of Threatened Species prepared by IUCN–The World Conservation Union.[31]

Plant Diversity Endangered

Kevin Eckerle

Recent assessments from IUCN–The World Conservation Union summarize some rather grim statistics on the current state of global plant diversity.[1] While the data are not representative in terms of the major plant taxonomic groups, geographic distribution, or ecosystems, they do provide a minimum estimate of the number of threatened plant species, offering a critical estimate of the overall condition of plant biodiversity.[2] Data from a 2004 assessment indicate that approximately 70 percent of the species assessed and 3 percent of all plant species are threatened with extinction (see Table 1), with almost 45 percent of the threatened plants listed as either endangered or, worse, critically endangered.[3] (See Figure 1.)

LINKS p. 92

Within the plant kingdom, the taxonomic group facing the greatest threat of extinction is the Gymnosperms—conifers, cycads, and ginkgoes. This is the plant group whose threatened status was most comprehensively assessed by IUCN in 2004, when it was determined that approximately one third of these species are threatened with extinction.[4] In a more recent study, 23 plant species, all of them conifers, were identified and categorized as facing imminent extinction because they were highly threatened and confined to a single site.[5]

Equally alarming is the loss of genetic diversity within individual domesticated plant species. Genetic variability is critical for maintaining the fitness and adaptability of any species. And since humans depend on a relatively small number of crop plants, a reduction in the genetic diversity of any single one represents a significant threat to human food security. While the reduction in the genetic diversity of crop plants is difficult to quantify, data indicate that there has been a significant loss in the genetic diversity of domesticated plant species.[6]

Plants are a major source of food and building materials. Estimates of the proportion of net primary productivity—the net amount of solar energy converted to plant organic matter through photosynthesis—that is appropriated by humans range widely, from 3 to 55 percent, with intermediate estimates of 20–32 percent.[7] These values are remarkably high, especially considering that the human population represents less than 1 percent of the total biomass of all heterotrophs—those organisms that, unlike most plants, cannot synthesize their own food and must derive nutrients by consuming other organic sources—on the planet.[8]

In addition, plants regulate atmospheric carbon dioxide, provide areas for recreation, and are essential for nutrient cycling.[9] The diversity in an ecosystem's plant community is positively related to diversity in the invertebrate and vertebrate community, overall productivity, and the long-term stability of an ecosystem.[10] As a result, any disruption in the underlying diversity of the plant community can have significant negative effects on the quantity and quality of the various services that an ecosystem is able to provide.

The major threats to plant diversity include invasive species, habitat destruction from increased agriculture, pollution from the use of nitrogen and phosphorus fertilizers, and global climate change.[11] While the magnitude of the stresses on plant diversity varies from low to very high, all of them are predicted to have greater impacts over the next 50–100 years.[12] As a result, the long-term predictions for global plant diversity are far from encouraging. The Millennium Ecosystem

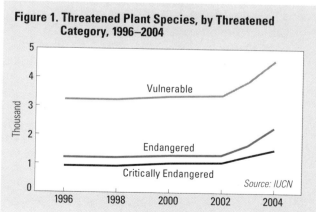

Figure 1. Threatened Plant Species, by Threatened Category, 1996–2004

Vulnerable

Endangered

Critically Endangered

Source: IUCN

Assessment—a four-year-long research effort involving 1,360 scientists—projects that 10–15 percent fewer vascular plant species than existed in 1970 will be supported by the habitat remaining in 2050.[13] Rates of vascular plant extinction are projected to be highest in warm mixed forests, savannas, scrub, tropical forests, and tropical woodlands.[14]

In a separate analysis of projected changes in the distributions of 1,350 European plant species in response to seven climate change scenarios, as many as 22 percent of the species were predicted to become critically endangered and 2 percent extinct by 2080. The predicted rate of species loss in this time (due to the loss of climatically suitable habitat) varied between 27 and 42 percent per 16 square kilometers. And the predicted rate of species turnover (change in species composition) varied between 45 and 63 percent per 16 square kilometers.[15]

These various assessments illustrate the challenge of significantly reducing the rate of loss of global biodiversity and, specifically, of plant diversity by 2010—which the Convention on Biological Diversity (CBD) set as its target year for such a reduction.[16] Yet as described in the Millennium Ecosystem Assessment, it is possible to achieve the 2010 target goals, at least for some components of biological diversity, in certain regions.[17] For instance, the rate of habitat loss in temperate terrestrial ecosystems is decreasing, and by preserving these habitats, the rate of loss of total biodiversity and plant diversity in these ecosystems can be reduced.[18]

But significant hurdles remain that may hinder progress toward the 2010 biodiversity targets. Most drivers leading to biodiversity loss are expected to increase in severity over both the short and the long term. There are also potential tradeoffs between the conservation of biological diversity and the goals of eradicating poverty and improving human health that the world's governments agreed to as Millennium Development Goals.[19] And the lag time between the implementation of specific conservation actions and our ability to quantify the effects of those actions on biological diversity can be several years or decades.[20] As a result, it seems highly unlikely that the CBD goal of significantly reducing the rate of biodiversity loss will be met by 2010, necessitating the development of a longer-term approach for combating the loss of Earth's plant and biological diversity.

Table 1. Threatened Plant Species, 1996–2004

Plant Species	Threatened Species					Species Evaluated in 2004	Species Described
	1996–98	2000	2002	2003	2004		
			(number)				
Mosses	—	80	80	80	80	93	15,000
Ferns	—	—	—	111	140	210	13,025
Gymnosperms (conifers, cycads, ginkgoes)	142	141	142	304	305	907	980
Angiosperms Monocotyledons (lilies, orchids, palms, grasses, and so on)	257	291	290	511	771	1,141	59,300
Dicotyledons (elms, maples, roses, buttercups, and so on)	4,929	5,099	5,202	5,768	7,025	9,473	199,350
Total	5,328	5,611	5,714	6,774	8,321	11,824	287,655

Source: IUCN.

Disappearing Mangroves Leave Coasts at Risk   *Zoë Chafe*

Over the last 25 years, 20 percent of the world's mangrove forests have been destroyed, leaving coastlines increasingly vulnerable.[1] Estimates from the U.N. Food and Agriculture Organization show that 15.2 million hectares of mangrove forests remain, down from 18.8 million hectares in 1980.[2] (See Table 1.) The rate of deforestation has slowed somewhat, however, from 185,000 hectares annually in the 1980s to 105,000 hectares a year since 2000.[3]

Mangrove trees require specific conditions to grow, restricting their geographic range. They flourish in the balmy waters of protected tropical shores and along some eastern subtropical coasts warmed by currents.[4] They are found exclusively in the intertidal zone, a habitat in which their roots are alternately submerged and uncovered. While they are able to grow in salt water, they also depend on nutrients from inland rivers and streams, which carry silt and eroded soils.[5] Mangroves are the only woody plants or grasses that can develop into a forest in this zone.[6] Facing adverse growing conditions, such as low soil oxygen, fluctuating salinity, and exposure to rough waves, they exhibit special adaptations—the ability to excrete excess salt from their leaves, for instance.[7]

LINKS pp. 42, 92, 102

Some 70 species of mangroves have been documented in 120 countries, though this diversity varies around the globe, peaking in humid Southeast Asia with 45 species in Indonesia and 44 species in Papua New Guinea.[8] Nearly half the area of the world's mangroves can be found in just five countries: Australia, Brazil, Indonesia, Mexico, and Nigeria.[9]

Mangrove forests house a wide variety of animals, including manatees, fishing cats, monitor lizards, sea turtles, oysters, crabs, and 500 bird species in Brazil alone.[10] Mangroves also shelter young fish before they move to nearby coral reefs.[11]

Protecting fish is only one of many important functions, or ecosystem services, that mangroves provide. With their extended roots, the trees can stabilize coasts, prevent erosion, and catch excess silt before it reaches sensitive offshore ecosystems.[12] They also serve as a natural filtration system, absorbing some pollutants. And mangrove forests have the potential to remove carbon dioxide from the air and act as a "carbon sink," mitigating climate change.[13]

Beyond ecological functions, mangroves play an important role in local economies. The wood can be converted to firewood, charcoal, logs, paper, and chipboard.[14] It is used in boat-making, favored by beekeepers for hives, and made into rot-resistant fish traps.[15] Mangrove poles have been exported from Tanzania and Kenya to the Middle East for more than 2,000 years.[16] Numerous medicines, such as contraceptives and aphrodisiacs, are derived from mangrove forests, as are dyes and insecticides.[17] Some tree species flower year-round and are used by beekeepers to ensure a constant supply of honey.[18] A single hectare of mangrove forest can produce $2,500 worth of non-timber forest products each year.[19] Mangroves themselves are becoming popular as ecotourism destinations.[20] And they are integral to local fisheries, with each hectare of intact forest in Indonesia generating about $9,900 of fisheries products annually.[21]

One important role of mangroves recently came to the world's attention: the trees protect coasts from strong waves. Mangroves blunted the destructive power of a 1999 cyclone that hit the Indian state of Orissa, as well as a 1991 cyclone and tidal wave that devastated Bangladesh.[22] In the wake of the 2004 Indian Ocean tsunami, scientists analyzing satellite images found that intact mangrove forests served as a "bioshield," protecting coasts and villages from forceful waves.[23] Anecdotal reports from most of the affected countries supported this finding. Along India's Andaman Islands, which have large stretches of virtually undisturbed mangrove forests, only 7 percent of the villages hit were severely damaged; where mangroves had been degraded, the share reached 80–100 percent.[24] Before the tsunami, human actions had decreased mangrove extent by more than 20 percent during the 20 years before 2000.[25]

Unfortunately, even as the importance of mangroves is being recognized, a variety of factors are contributing to their degradation and disappearance. In many places, the forests are cut to make way for aquaculture: shrimp farms and fishponds. In the Philippines, national development policy during the 1950s led to the degradation of thousands of hectares of mangroves, many of which were turned into ponds for milkfish, a variety favored by local people.[26] As a result, only 20 percent of the half-million hectares of mangroves that existed in 1900 now remain, though the area covered by fishponds has quadrupled to 230,000 hectares.[27] In East Africa, mangroves are cleared for the production of salt.[28] Irresponsible tourism development and coastal construction also contribute to degradation.

Luckily, a few innovative projects are saving mangroves. Spurred by lessons from the 2004 tsunami, several countries are working to restore these forests. In Indonesia, a U.N. Development Programme official reported that after the disaster the first priority was to replant 1,600 kilometers of destroyed mangroves along the Sumatran coast, to help local fisheries recover.[29] In India, the government of Maharashtra state has outlawed mangrove destruction, prohibited construction within 50 meters of the forests, and declared all mangroves on government land "protected forests."[30] And in Malaysia, Prime Minister Abdullah Badawi has discouraged the degradation of mangroves for the sake of development and called for replanting in damaged areas.[31]

IUCN–The World Conservation Union recently announced a $45-million project to replant mangroves and restore livelihoods in five countries severely affected by the tsunami.[32] Mangroves and other coastal trees can be replanted at a cost of $150–2,000 per hectare in tsunami-affected areas.[33] In Pakistan, the World Wide Fund for Nature (WWF) International is working with community-based organizations and local fishers to rehabilitate mangrove forests on the Balochistan Coast.[34] And in Indonesia, the Wetlands International Indonesia Programme has been organizing replanting projects since 1998, employing villagers to collect and plant wild seedlings in return for small business loans.[35] If 70 percent of the mangroves

Year	Estimate
	(million hectares)
1980	18.8
1990	16.9
2000	15.7
2005	15.2

Table 1. Global Mangrove Area Estimates

Source: FAO.

are still alive after five years, the villagers need not pay back the loans.[36]

WWF has announced plans to reestablish shrimp farms in areas affected by the 2004 Indian Ocean tsunami, a move that has been questioned by several organizations because shrimp farming led to much of the mangrove destruction that left coasts more vulnerable during the tsunami.[37]

Because shrimp and fish farming can be so harmful to mangroves, several groups have established innovative projects to give consumers more-responsible seafood choices. A scientist in the Philippines is advocating that fishponds be established outside of mangroves or that sustainable forest cultivation techniques be used within intact mangrove forests.[38] And in Belize, WWF is working with prawn farmers to help them produce and certify prawns that have been raised with environmentally sensitive aquaculture techniques.[39]

While several seafood certification programs purport to address mangrove degradation, there is an immediate need to integrate more-stringent standards into certification criteria. Alfredo Quarto of the Mangrove Action Project has warned that many widely recognized labels, such as the one Wal-Mart endorsed in 2006, fail to adequately prevent further destruction of mangrove forests for seafood and shrimp production.[40] With so few of these forests remaining, it is vital that mangroves not be further harmed by the rising demand for such products.

Deforestation Continues

Gary Gardner

Global forested area contracted by some 36.6 million hectares, or just under 1 percent, between 2000 and 2005, a continuation of the decades-long trend of forest loss in much of the world, according to the *2005 Global Forest Resource Assessment* by the Food and Agriculture Organization (FAO).[1] (See Table 1.) The loss is a net calculation: total losses were actually some 65 million hectares over the five-year period, but these were partially offset by expansion of plantation forests and the regrowth of some natural forests.[2] The net contraction in the first five years of the new century is about 19 percent less than the shrinkage during comparable periods in the 1990s.[3]

LINKS pp. 92, 100

The largest losses were in Africa, which shed 3.2 percent of its forested area between 2000 and 2005, and South America, which lost 2.5 percent.[4] Eight nations in Africa registered losses of more than 10 percent over the five-year period, a rate of contraction that would eliminate their forests within a few decades if unchecked.[5] In South America, every country except Ecuador lost forested area in the past half-decade. Brazil, home to the Amazon forest, lost 3.2 percent of its total area, an acceleration of deforestation since the 1990s.[6]

Every country in Europe reported either no change in forested area or a slight increase, with total expansion between 2000 and 2005 amounting to 0.33 percent of the continent's forested land.[7] Asia also registered a net increase, of 0.88 percent; most of this growth came in China, where an ambitious reforestation program reportedly boosted the country's forested area by some 11 percent.[8] Illegal logging is especially problematic in Asia, however, and accounting for it could dampen the net reforestation trend in the region.[9]

The most recent forest losses cap a long history of deforestation at the global level. In 1997 the World Resources Institute calculated that the world had lost some 46 percent of its forests in the past 8,000 years and that most of the trees had been cut down in the last three decades of the twentieth century.[10]

The FAO assessment reported that only 36

Table 1. Change in Extent of Forest, 2000–05

Region	Change in Area (thousand hectares)	Change in Area (percent)
South America	−21,256	−2.5
Africa	−20,201	−3.2
Oceania	−1,780	−0.86
Central America and Caribbean	−1,158	−3.9
North America	−507	−0.07
Europe	3,303	0.33
Asia	5,015	0.88
World	−36,583	−1.0

Source: FAO.

percent of the world's forests are primary forests—those that are largely undisturbed by direct human activities.[11] Roughly half of global forested area is "modified natural forest," according to FAO.[12] Plantation forests—those that are seeded, typically with fast-growing species that produce regular harvests for commercial use—are increasingly common and now account for 3 percent of the world's forested area.[13] But these are typically home to fewer species and are less robust than natural forests.

Deforestation occurs principally when people clear forests for agriculture or building or when they harvest wood for fuel or lumber. But these direct causes are in turn driven by more indirect factors such as poverty, economic growth, government policies, technological change, and cultural factors. Agricultural expansion is the single most important cause of forest loss, involved in 96 percent of cases in a 2001 study of 152 cases of deforestation going back to 1880.[14] Yet it was rarely the sole cause; it was often connected with timber harvesting and road building.

The role of agriculture, especially cattle-raising, in deforestation is especially clear in the Amazon. Between 1990 and 2002, 80 percent of the increase in cattle populations in Brazil came in the Amazon.[15] Cattle populations more than doubled there, to some 57 million head.[16] The expanded herds were accommodated by the opening up of new pastureland, at the expense of forest area. The process was driven by low land prices and a devalued Brazilian currency, which made Brazilian beef cheap and stimulated a quintupling of beef exports between 1997 and 2003.[17]

Beyond their commercial value, forests provide myriad ecological services, including habitat for diverse species, erosion control, and regulation of the hydrological cycle.[18] They are also an important sponge for atmospheric carbon and therefore vital in the effort to stabilize the climate. As forested area contracted between 1990 and 2005, the carbon storage capacity of the world's forests declined by more than 5 percent.[19] On the other hand, forests can be a leading source of carbon emissions when they are cut—because rotting branches, trees, and other debris give off carbon dioxide. Indeed, FAO concludes that deforestation accounts for 25 percent of the annual emissions of carbon caused by human activity.[20]

Even when forests are restored, they may require extended periods to recover their full ecological productivity. Researchers at Ohio State University in 2005 found that the carbon content of regrown forests—some of which had been in recovery for 70 years—was just one half the amount stored in nearby forest stands that had never been cut.[21] The scientists were unsure how long it would take for the newer stands to achieve the productivity of the original trees.

So valuable is the carbon capture service provided by forests that it may well trump their value in supplying timber or other products. According to a November 2005 analysis, carbon storage by the tropical forests of 10 developing nations was worth $1.1 trillion (based on $20 per ton of carbon dioxide, its rough value in carbon markets at the time).[22] In contrast, the value of roundwood, fuelwood, and non-wood forest products from these nations is easily an order of magnitude lower.[23] Adding other non-marketed ecological services to the mix would greatly increase forests' monetary value.

The Coalition for Rainforest Nations, a

Table 2. Top 10 FSC-Certified Countries

Country	Certified Area, January 2006	Increase Over Area Reported in 2005
	(million hectares)	(percent)
Canada	15.39	252
Sweden	10.42	3
Russia	6.70	216
Poland	6.25	1
United States	5.62	5
Brazil	3.53	34
Bolivia	2.20	n.a.
Croatia	1.99	no change
Latvia	1.69	no change
United Kingdom	1.60	32

Source: Forest Stewardship Council.

group of developing countries with extensive tracts of tropical forest, proposed at climate talks in Montreal in December 2005 that its member nations be paid for not cutting their forests.[24] They argued that they should be compensated for the losses incurred by forgoing the cutting of forests and that only strong economic incentives for protection will prevent forests from being cut.[25] Under the Kyoto Protocol, credits are given when new forests are planted but not when standing forests are preserved. The proposal was referred to future climate meetings for review and possible adoption.

The area of forest that is certified as sustainably managed continues to expand globally.[26] (See Table 2.) The Forest Stewardship Council (FSC) reports that certified area reached 68 million hectares in 66 countries in January 2006, a 45-percent increase in just one year.[27] Still, total certified area amounts to less than 2 percent of total global forested area.[28] In Europe, certified area represents 3.5 percent of total forested area, in North America 3.2 percent, and in Central America and the Caribbean, 2.6 percent.[29] Some individual countries stand out, however, in their certification efforts: 93 percent of Croatia's forests, 68 percent of Poland's forests, and 38 percent of Sweden's forests are FSC-certified.[30]

Groundwater Overdraft Problem Persists
Yingling Liu

Ready access to the technologies of drilling deep wells and increased pumping capacity have led to groundwater depletion in many parts of the world, a condition under which groundwater is being pumped out faster than nature can recharge it. This "groundwater deficit" has widespread social and economic impacts, with subsidence—the gradual settling or sudden sinking of the land surface—and seawater intrusion being two particularly important ones.

Rapid and largely uncontrolled expansion in groundwater exploitation has been rampant for

LINKS pp. 22, 24

decades due to rising water demand. Overexploitation of groundwater occurred in many industrial nations during 1950–75 and in most parts of the developing world during 1970–90.[1]

Groundwater overdraft and aquifer depletion are now serious problems in the world's most intensive agricultural and urbanized regions. Even with concerted efforts to eliminate overdrafting of groundwater, withdrawals are expected to increase 18 percent between 1995 and 2025.[2] (See Table 1.) In the United States, groundwater depletion has been a concern in the Southwest and High Plains for many years. It has also been observed in the Atlantic coastal plain, Gulf Coast plain, west-central Florida, the Chicago-Milwaukee area, and the Pacific Northwest.[3] The High Plains Ogallala aquifer is being depleted eight times faster than nature can replenish it. And the water table under California's San Joaquin Valley has dropped nearly 10 meters in some spots within the last 50 years.[4]

In Mexico, 130 out of 459 aquifers are considered overexploited or threatened with overexploitation.[5] The aquifer that supports the 1.5 million people living in the semiarid Ciudad Juárez/El Paso region is being drawn down rapidly and is expected to be depleted in a mere 20 years.[6]

Current depletion of Africa's non-recharging aquifers is estimated at 10 billion cubic meters a year.[7] The Arabian Peninsula fares no better, with groundwater use at nearly three times greater than recharge.[8] At the current rate of extraction, Saudi Arabia is racing toward total

depletion in the next 50 years, and Israel's extraction has exceeded replacement by 2.5 billion cubic meters over the last 25 years.[9]

The two most populous nations in the world, China and India, are suffering from significant groundwater depletion as well. From 1995 to 2004, India observed a significant decline in the water table during the pre-monsoon period, especially in the more arid western and southern states underlain by weathered bedrock aquifers. The annual decrease amounted to 20 centimeters per year.[10]

The shallow aquifer under the North China Plain, which harbors that nation's major agricultural bases and urban centers, including Beijing, has experienced a water table decline of more than 15 meters in the past 30 years.[11] The water table beneath Beijing and the surrounding Hebei Province has dropped 20–40 meters.[12] The whole region has seen an area of 76,732 square kilometers of deep groundwater table drop below sea level, accounting for 55 percent of the North China Plain.[13]

Large-scale groundwater withdrawal can lead to a consolidation of aquifers, causing land subsidence. This in turn causes widespread damage to urban centers—undermining buildings and rupturing roadways, water lines, sewer systems, and other infrastructure. It also jeopardizes sunken areas with flooding and even permanent inundation. More than 80 percent of the subsidence in the United States is related to the withdrawal of groundwater.[14]

Parts of the United States suffering from subsidence include California's San Joaquin Valley, the Houston-Galveston area in Texas, Baton Rouge in Lousiana, and the Phoenix area in Arizona.[15] Land subsidence caused by groundwater extraction has been measured at as much as 2.4 meters around the San Francisco Bay; in the vicinity of the City of Los Angeles, it occurred at an annual rate of 0.7 meters over several decades.[16] In downtown Mexico City, the land has sunk more than 7 meters since 1900.[17] Present-day rates vary from a millimeter or two per year to more than 40 centimeters a year in the southeastern portion of Mexico City.[18]

In China, by 2003 more than 50 cities had

seen land subsidence, with the area affected reaching 94,000 square kilometers.[19] Yangtze Delta, the North China Plain, and Fenwei Basin were the worst cases; Shanghai and parts of Jiangsu and Zhejiang provinces have seen almost 3 meters accumulated subsidence.[20]

Coastal aquifers, on which many of the world's largest cities and most concentrated industrial bases are located, are suffering from seawater intrusion as well. When groundwater extraction in these areas reaches a certain point, seawater can be pulled into freshwater aquifers. The resulting contaminated groundwater is virtually unusable. To many fast-urbanizing areas that depend largely on groundwater for drinking and agriculture, the saline water not only degrades their soil, it poses serious threats to their drinking water supply.

In the United States, seawater intrusion is especially a concern along the Atlantic coast.[21] It occurs in coastal counties in New Jersey, South Carolina, Georgia, and Florida, as well as in the Tampa-St. Petersburg area and Gulf of Mexico coastal aquifers.[22]

Seawater intrusion has been a serious concern to India and China as well—two nations already facing a freshwater crisis. Widespread coastal salinity has been observed in India's Mangrol-Chorwad areas and coastal Saurashtra of Gujarat, the minjur area in Tamil Nadu, the Pondicherry coast, parts of Orissa, and Andhra Pradesh and Kerala coasts.[23] In China, coastal provinces and regions including Liaoning, Hebei, Shandong, Guangxi, and Hainan are suffering the most.[24] Seawater intrusion in coastal regions around Laizhou Bay have threatened drinking water safety for more than 400,000

Table 1. Groundwater Withdrawal in 1995 and under Business-as-Usual and Low Groundwater Pumping Scenarios, 2021–25

Region/Country	Baseline in 1995	Projections for 2021–25	
		Business as Usual	Low Groundwater Pumping
		(cubic kilometers)	
Asia	1,953	2,420	2,286
China	679	844	813
Southeast Asia	203	278	278
South Asia, excluding India	353	416	392
India	674	822	743
Latin America	298	402	402
Sub-Saharan Africa	128	207	208
West Asia/North Africa	236	294	270
Industrial countries	1,144	1,272	1,266
Developing countries	2,762	3,481	3,324
World	3,906	4,752	4,590

Note: 2021–25 data are annual averages.
Source: Rosegrant, Cai, and Cline.

people, rendered more than 8,000 agricultural wells unusable, deprived 40,000 hectares of arable land of irrigation, and caused an annual reduction of grain production amounting to 300,000 tons.[25]

To counter these problems of groundwater depletion, governments are turning to a variety of remedies. The United States has developed conjunctive surface water-groundwater management, widely known as groundwater banking—a scheme whereby carefully calculated amounts of streamflow are diverted during wet years and used to recharge vast alluvial aquifers.[26] China has started to build an ambitious south-north water diversion project to meet fast-growing water demands and ease groundwater overdraft in the north. The project has three water diversion routes, connecting the Yangtze with the largest rivers in the north—the Yellow, the Huaihe, and the Haihe.[27] And India has been using water harvesting—simple technologies used to capture and store water before it can flow away—to inject rainwater into aquifers to raise its declining water table.[28]

Reducing Mercury Pollution

Linda Greer and Michael Bender

Governments across the world increasingly warn people to restrict their intakes of certain types of fish to avoid excess exposure to mercury, a potent toxic metal that interferes with brain functions and the nervous system. Many populations are further exposed to mercury from a variety of other sources, including their work (very high exposure in some circumstances), consumer products, waste disposal, and even health care products (such as dental tooth fillings, cosmetic preparations, and some vaccines).

LINKS p. 32

The populations most vulnerable to mercury's toxic effects are pregnant women (because it affects fetuses) and children. Even in low doses, mercury exposure may influence a child's neurological development, affecting attention span, fine-motor function, language, visual-spatial abilities (such as drawing), and verbal memory.[1] In adults, chronic mercury poisoning is also a serious health threat and can cause memory loss, tremors, vision loss, and numbness of the fingers and toes and can contribute to heart disease among other problems.[2]

Mercury is a classic global pollutant: when released from a source in one country, it readily disperses around the world, often falling far from its source of release and entering distant food supplies. These characteristics have led to surprising and disturbingly high concentrations in places with no significant local mercury pollution sources at all. The Arctic region, in particular, is a global mercury hotspot, acting as a giant "sink" for the pollutant circulating in Earth's atmosphere. People there are some of the most highly contaminated on Earth.[3] Mercury concentrations are also extremely high in top predators of global ecosystems, such as seals and polar bears in the Arctic and toothed whales, seabirds, and loons.[4]

Mercury emissions have risen three- to fivefold over the past two centuries due to anthropomorphic releases.[5] The last 40 years have seen a significant increase in emissions from coal combustion, although that has been offset to some degree by a reduction in industrial uses of mercury worldwide, from more than 9,000 tons per year in the 1960s to less than 4,000 tons annually since 2000.[6] The industrial decline has occurred largely because various countries have made conscious decisions to decrease mercury use by reducing or eliminating it in products such as batteries and paints and by converting industrial processes, such as chlor-alkali plants, to mercury-free technology.[7]

Unfortunately, however, largely as a result of these changes, a flood of surplus mercury has entered markets in the developing world, often into uncontrolled or poorly controlled uses. The resulting uses and releases pose large local risks to human health and the environment as well as contributing substantially to the total quantities of mercury circulating worldwide.

The most important remaining uses of mercury are for product manufacture (such as batteries, measuring devices, and switches), the chlor-alkali industry (which manufactures chlorine and caustic soda from brine), and artisanal and small-scale gold mining, which together account for up to two thirds of the global total.[8] (See Figure 1.) Eighty percent of mercury is currently used in developing countries, particularly in Asia.[9] China and India appear responsible for nearly 50 percent of global mercury demand.[10] The European Union has been the major global mercury supplier, in 2000 shipping nearly 1,000 tons to South Asia and the East Asia/Pacific regions alone, meeting nearly half of those regions' needs.[11] China, however, has recently met all its needs through its own mercury mines.[12]

Fortunately, economically viable alternatives to mercury are available for nearly every application, as are control technologies that can reduce or eliminate releases from the largest sources of pollution. These options have allowed industrial nations to reduce mercury use and releases, as well as occupational exposures, substantially. Many governments in the developing world have not made mercury reductions a priority, however, and do not track mercury use within their borders. Without financial assistance and international agreements, they are unlikely to move forward on any reduction initiatives.

The dynamics of the mercury market make

it crucially important that a globally focused mercury reduction strategy ratchet down supply and demand simultaneously. Only with a coordinated global strategy can governments ensure that steps taken to reduce demand in some locations do not flood the market with excess mercury supplies elsewhere, which would invite mismanagement. Similarly, a global strategy will ensure that a plummet in supply does not trigger new mining to meet unsatisfied demand.

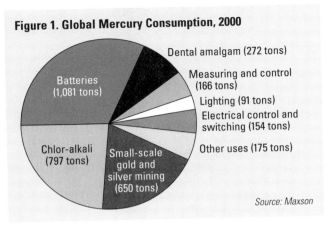

Figure 1. Global Mercury Consumption, 2000

Batteries (1,081 tons)

Dental amalgam (272 tons)

Measuring and control (166 tons)

Lighting (91 tons)

Electrical control and switching (154 tons)

Other uses (175 tons)

Chlor-alkali (797 tons)

Small-scale gold and silver mining (650 tons)

Source: Maxson

Mercury trade data indicate a relatively small number of supply and demand sources of mercury worldwide. This creates opportunities to achieve large global reductions through progress in a handful of large uses. Indeed, focusing on meeting only three of the major global sources of mercury demand with readily available alternatives—batteries, mercury cell chlor-alkali plants, and switches/measuring devices—could reduce global mercury demand 50 percent by 2010 and 75 percent by 2015.[13] Global mercury supply can also be dramatically reduced on a parallel track through policies that close the handful of remaining virgin mercury mines in the world (currently found only in Algeria, China, Kyrgyzstan, and Spain), accompanied by commitments to store excess mercury surpluses after phasing out high-volume uses in the industrial world.

In addition, since the global need for energy is expanding so rapidly, an aggressive mercury control strategy for new and existing coal-fired power plants is crucial in order to reduce the global mercury load adequately. To achieve this, coordinated international action must focus on installing the best available technology for mercury emission controls for major coal-fired power plants.

While global progress on mercury reductions is slow, it is in fact under way. The European Union, for example, has developed a proactive and coordinated mercury reduction strategy that restricts exports of surplus mercury in the coming years as well as shutting down mercury mining in Spain.[14] A mercury use inventory is under way in China, which will illuminate opportunities to reduce demand.[15]

Clearly, reductions in global mercury supply and demand can initially be accomplished through a variety of mechanisms, including voluntary actions, domestic legislation, and aid packages, as well as through national, bilateral, and regional arrangements. Ultimately, however, reductions will undoubtedly require a binding, global agreement. There are alternatives to mercury, but there is no alternative to international determination, cooperation, and action. International action to address the global mercury problem is essential to human health in all parts of the world—from New York to London, Beijing to Johannesburg, and even within the Arctic Circle.

Economy and Social Features

Eight-year old slum-dweller breaks the ribbon first in a school race, Dhaka, Bangladesh

▶ Regional Disparities in Quality of Life Persist

▶ Language Diversity Declining

▶ Slums Grow as Urban Poverty Escalates

▶ Action Needed on Water and Sanitation

▶ Car-sharing Continues to Gain Momentum

▶ Obesity Reaches Epidemic Levels

▶ Corporate Responsibility Reports Take Root

▶ Nanotechnology Takes Off

For data and analysis on these economic and social issues and on literacy, unemployment, and asthma, go to www.worldwatch.org/vsonline.

Regional Disparities in Quality of Life Persist

Peter Stair

Although true measures of human quality of life are too subjective to quantify, proxy measures indicate that some regions of the world have become more prosperous while others have become more troubled. The Human Development Index—a composite of life expectancy, literacy, school enrollment, and per capita economic output—shows that conditions have been improving in Western Europe, the Americas, and East Asia.[1] (See Figure 1.) But the disorder brought by the fall of communism has caused living conditions to deteriorate in many East European and former Soviet states, and both wars and the HIV/AIDS epidemic have significantly reduced the quality of life in sub-Saharan Africa.[2]

pp. 52, 74, 76, 92, 114, 116

Overall, more people are acquiring the basics they need for survival. In developing countries, the number of people in absolute poverty—defined as an income of $1 a day or less—declined from 28 percent in 1990 to less than 21 percent in 2003, a total of 230 million fewer impoverished people.[3] The proportion of people who are chronically undernourished has dropped since the 1970s, and between 1990 and 2002 there was an absolute decline of 9 million in this category in developing countries.[4] In the same period, the proportion with access to better sanitation jumped from 43 to 58 percent, while the share with cleaner water increased from 75 to 83 percent.[5]

The average human life span has continued to lengthen, increasing by two years in developing and industrial countries between 1990 and 2002.[6] This is due largely to improvements in prenatal and infant child care, which reduced the number of deaths among children younger than 5 by 2 million during these years.[7] The Global Alliance for Vaccines estimates that vaccinations have lowered the number of deaths caused by viruses, especially hepatitis B, by 1.7 million since 2001, while monitored treatment of tuberculosis has cut its prevalence by more than 20 percent since 1990.[8]

More people are able to participate peacefully in their societies. The number of violent conflicts has declined since 1990, as the fighting decreased in places such as East Timor, Afghanistan, El Salvador, Sierra Leone, and Liberia.[9] Meanwhile, more people have learned to read, a proxy measure for their ability to participate in economic and governmental systems. The literacy rate increased in developing countries from 70 percent to 76 percent over the last decade.[10] And the gender gap in education has also been shrinking—in the Caribbean, Western Asia, Europe, and the United States, women outnumber men in higher education.[11]

The share of countries considered "democratic" by the Center for International Development and Conflict Management went from 39 percent in 1990 to 55 percent in 2003.[12] Over the same period, the share of national governments considered "autocratic" declined from 39 percent to 18 percent.[13] Today an additional 1.4 billion people live in countries that have regular multiparty elections, and women are slowly gaining rights in this area as well.[14]

All these trends paint an encouraging picture of the human condition, but they mask huge regional disparities. Much of the progress in poverty, hunger, and life span occurred in East Asia, where economic reforms and growth have spurred improvements. By contrast, droughts, warfare, and the HIV/AIDS plague have worsened living conditions significantly in sub-Saharan Africa. To illustrate the difference: between 1990 and 2002 the number of hungry people in East Asia declined by 47 million, while it increased by 34 million in sub-Saharan Africa.[15]

In former Eastern bloc countries, the transition from communism has been difficult. The expected life span of a Russian dropped from 70 years in the mid-1980s to 59 in 2002, due to a rise in poverty, a resurgence of diseases such as tuberculosis and diphtheria, a increase in the rate of alcoholism, and the spread of HIV/AIDS.[16] In former Soviet countries, the rate of extreme poverty jumped from 0.4 percent to 5.3 percent between 1990 and 2001.[17]

Globally, there is evidence that recent gains in poverty and health are slowing. Extreme poverty is declining at just one fifth the rate of the 1980s, and between 1997 and 2002 the absolute number of hungry people increased.[18] The World Health Organization (WHO) has identified six diseases that are reemerging due to changes in climatic and sanitary conditions as well as an increase in international travel: influenza, diphtheria, cholera, dengue fever, yellow fever, and bubonic plague.[19]

Other global trends have also continued to worsen. Economic inequality has been increasing in most countries—80 percent of the world lives in countries where economic disparity increased between 1990 and 2003.[20] And suicides—one very personal measure of human well-being—are becoming more common. According to WHO, the worldwide suicide rate among men rose between 1950 and 2000 from approximately 17 per 100,000 people to about 30.[21]

Although most industrial countries have already achieved near-universal literacy, long life spans, stable economies, and democratic governments, they still have pockets of entrenched poverty. In the United States and Europe, minority groups and immigrants in particular often live in conditions comparable to those in developing countries.[22]

There is also evidence that continued economic growth, usually considered the linchpin of human development in the developing world, is bringing diminishing returns in the industrial world. Studies show that after reaching a certain level of income, additional wealth no longer improves peoples' sense of their happiness.[23] Likewise, an abundance of food and physical ease has contributed to a rise in obesity, type-II diabetes, and cardiovascular disease—health problems that may even cause life expectancy in industrial countries to decline.[24]

Nonetheless, the World Database of Happiness—a collection of surveys that asked people in industrial countries to rate their happiness on a scale from 0 to 10—indicates that personal happiness slowly increased between 1973 and 2004 in most countries surveyed, particularly

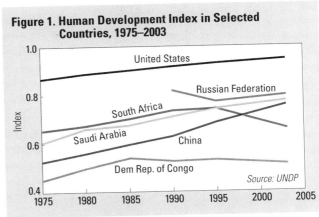

Figure 1. Human Development Index in Selected Countries, 1975–2003

Source: UNDP

Italy and Denmark.[25] Only Belgians reported a notably lower level of happiness.[26]

Continued improvements in human well-being will be limited by the planet's ecological carrying capacity. For example, changes to the climate brought by the clearing of land and the burning of fossil fuels threaten not only to submerge coastal human settlements but also to expand the prevalence of tropical diseases and reduce overall agricultural harvests. The U.N. Food and Agriculture Organization has concluded that global warming may "drastically" increase the number of hungry people.[27] As reiterated recently by the Millennium Ecosystem Assessment, it is becoming clear is that as human demands butt up against global limits, human and environmental well-being will be increasingly tied together.[28]

Language Diagram Declining

Language Diversity Declining

Courtney Berner

Linguists estimate that 10,000 years ago there were approximately 5–10 million people on Earth and up to 12,000 spoken languages.[1] The current global population is more than 6 billion, yet the number of unique languages has shrunk to fewer than 7,000.[2] Some experts estimate that we lose one language every month, while others peg the loss at one every two weeks.[3] At this rate, 100 years from now there could be only 2,500 languages left. Worse, some experts maintain that 90 percent of the world's languages will vanish or be replaced by dominant languages by the end of this century.[4]

While the mental processes of knowing and learning about a person's environment may be fundamentally the same regardless of the language that person thinks in, languages are full of unique cultural knowledge that facilitates different ways of understanding and discussing the world.[5] In addition to the cultural, historical, and purely linguistic importance of languages, native tongues are also often rich in knowledge of the local flora and fauna and their traditional medicinal uses. Some notable drugs—aspirin, codeine, ipecac, pseudophedrine, and quinine, to name a few—have been developed from knowledge derived from languages that are now endangered or extinct.[6]

More than half of the world's 7,000 languages are endangered, and nearly 550 languages are spoken fluently by fewer than 100 people, increasing the likelihood that they will disappear quickly.[7] Of these, 516 are considered nearly extinct.[8] A language is classified as nearly extinct when the speaker population is fewer than 50 or when the number of speakers represents a very small fraction of an ethnic group.[9]

Fifty percent of the world's languages are losing speakers, and 95 percent of them are spoken by just 6 percent of people worldwide.[10] Only 347 languages, about 5 percent of the total, are spoken by more than a million people.[11]

Asia is home to the highest share of languages worldwide, followed by Africa and the Pacific.[12] (See Table 1.) Papua New Guinea and Indonesia contain the greatest number of languages, with 820 and 742 respectively.[13] They are followed by Nigeria with 516, India

Table 1. Number of Languages and Share of World Total, by Region

Region	Number of Languages	Share of Total
		(percent)
Asia	2,269	32.8
Africa	2,092	30.3
Pacific	1,310	19.0
Americas	1,002	14.5
Europe	239	3.5
World	6,912	100.0

Source: Summer Institute of Linguistics.

with 427, the United States with 311, and Mexico, Cameroon, and Australia with just under 300 each.[14] Half of the world's languages are spoken in only eight countries.[15]

The same conditions that breed biological diversity—isolation and variations in climate, terrain, and ecosystems—also produce language-rich environments.[16] Not surprisingly, the countries with the greatest number of languages are also the ones losing languages the fastest. Brazil, which is suffering the loss of both biological and linguistic diversity, is home to 30 nearly extinct languages.[17] The United States has close to 70 nearly extinct languages, and Australia leads with 188.[18] It is estimated that 90 percent of the languages spoken by Australia's Aboriginal peoples will perish with the current generation.[19] In Africa, approximately one third of tribal languages are endangered.[20] Saving them will be very difficult, since 80 percent of African languages have no written record.[21]

The disappearance of languages is not easily understood or documented because it is often the result of complicated social, cultural, and historical processes that typically unfold within small speech communities during periods of political and socioeconomic change.[22] The death of a language is most commonly caused by bans on regional languages, infectious diseases, wars, migration, and cultural assimilation.[23] Most languages disappear because their speakers voluntarily abandon them.[24] When a community finds that its ability to survive and advance economically is improved by the use of another language, for example, people there stop using their native tongue or teaching it to their children.[25]

While a language may be endangered because the number of speakers drops, the true measure of endangerment is the range of its context of use within a community of speakers. If the number of situations in which a language is useful is decreasing (if it is used only in religious ceremonies, for instance, or for bedtime stories), then it is more likely to fall out of use.[26] To be preserved, a language must be used regularly in a community of speakers who pass that language on to their children.

Nearly 1 billion people use Mandarin Chinese as their first language.[27] (See Figure 1.) English, however, is the most commonly spoken second language and the lingua franca in the international business, media, scientific, and academic worlds.[28] English language classes are soaring in popularity, particularly in China. It was estimated that by 2000 more than 1 billion people worldwide were learning English, and the number of English-speaking residents of China is growing by 20 million every year.[29] With the 2008 Beijing Olympic Games approaching, city officials are encouraging residents to learn English by providing learning opportunities and setting proficiency targets for certain segments of the population. For example, 80 percent of police officers under the age of 40 should be capable of passing a basic oral English exam.[30] In the United States, Chinese is expected to overtake French and German in popularity in school classrooms by 2015.[31]

Although only 10 percent of the world's languages are represented on the World Wide Web, the Internet offers a relatively inexpensive way for speakers and advocates of endangered languages to attract attention and possibly new speakers.[32] The problem of unequal access to the Internet, however, must be solved

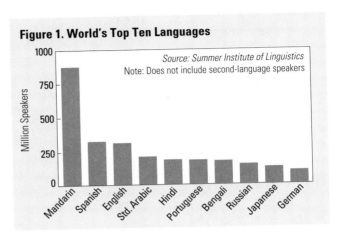

Figure 1. World's Top Ten Languages

Source: Summer Institute of Linguistics
Note: Does not include second-language speakers

if this technology is going to have any positive impact on language conservation.[33]

While languages are disappearing quickly, in a few places traditional languages continue to be used and preserved. For more than 20 years, Kahnawa:ke, a small community near Montreal in Canada, has supported the use of the Mohawk language through immersion schools, church services, and even a local radio station.[34] In January 2002, it became the first Mohawk territory to enact its own language law.[35] And in June 2005, the European Union (EU) added Irish Gaelic to its list of official languages.[36] About 30 people will be hired to translate legislation and speeches at EU meetings, all to help preserve Ireland's native tongue.[37]

The work of academics, governments, communities, and nonprofit groups is helping to slow the disappearance of some endangered languages. In July 2005, UNESCO's Endangered Languages Programme, which promotes and safeguards linguistic diversity, launched a Register of Good Practices in Language Preservation.[38] This provides a place where the results of language protection projects can be reported and made publicly available in order to give future preservation projects a head start. Similarly, the U.S. National Science Foundation and the National Endowment for the Humanities recently established a $4.4-million Documenting Endangered Languages project, the first of its kind.[39]

Slums Grow as Urban Poverty Escalates   Molly O'Meara Sheehan

Soon half of humanity will live in urban areas, where insecure, disease-filled slums are swelling. Demographers project that in 2007 world population will tip from being more rural to being more urban.[1] Nearly all population growth between 2000 and 2030 is expected in urban areas of the developing world.[2]

In *State of the World's Cities 2006/7*, UN-HABITAT estimates that 1 billion individuals, one in every three urbanites, live in "slums"—areas where people cannot secure one or more

pp. 78, 110, 116

of life's basic necessities: clean water, sanitation, sufficient living space, durable housing, or "secure tenure," which includes freedom from forced eviction.[3] It might not be right to apply the harsh term "slum" to the homes of hard-working families in vastly different locales, but it is essential for governments to count poor people and to consider the economic and political benefits that follow when slums improve and people reach their full potential.

"Slum" is one of the many loaded words used to describe wretched urban conditions. Rent-paying tenants, not true "squatters," often occupy "squatter settlements" that form when people build on land they do not own. Law-abiding citizens pay landlords who supply housing that governments deem "illegal" or "informal." "Slum," coined in post–Industrial Revolution London, does not reflect the disparities among and between the "hoods" of New York, *villas miserias* of Buenos Aires, *bidonvilles* of Abidjan, *zopadpattis* of Mumbai, and countless places that each go by multiple names.[4] But "slum" serves as shorthand for urban insecurity.

News crews descended on a few slums in wealthy nations in 2005. Lacking adequate public transport or bus fare, tens of thousands of New Orleans' poorest citizens who were trapped in Hurricane Katrina's path in August struggled to survive the storm and its aftermath without enough clean water, food, sanitation, or security.[5] While some of the city's wealthiest protected their property with Israeli commandos in old Soviet assault helicopters, scenes of despair and death in the Louisiana Superdome showed the horrors of poverty compounded by

inept government.[6] A bus driver in Somalia told a reporter: "New Orleans looks like Mogadishu when the war started."[7]

Images of Paris burning dominated the television news in October and November, when young people from low-income suburbs set cars and buildings ablaze after two teenagers of African descent died from electrocution in an electrical substation, where they had fled to avoid police.[8] The riots showed the rage of young people plagued by discrimination, inadequate education and employment prospects, and fire-prone housing.[9] Between April and September, three fires in cheap hotels and apartment buildings in Paris killed 48 African immigrants, mostly children.[10]

More than 90 percent of slums are in Africa, Asia, and Latin America, and pictures of them are rarely televised.[11] With the 2005 film *The Constant Gardener*, about Nairobi, and the 2002 Rio epic *City of God*, director Fernando Meirelles gave a global audience a rare glimpse of what life is like for a third of the world's urban residents in slums.

The world's fastest urbanization, in sub-Saharan Africa, is virtually synonymous with slum growth: the region's urban population is rising annually by 4.58 percent while slums are expanding by 4.53 percent.[12] (See Figure 1.) Slum growth also comes close to urban growth in Western Asia and, to a lesser extent, Southern Asia.[13] In Latin America, the most urbanized developing region, slum growth has slowed as poverty has become largely urban: 134 million of the poorest 211 million people live in urban areas.[14] Only in Northern Africa, where governments have invested in urban water and sanitation, have slums declined as cities have grown.[15]

Throughout history, people have fled rural poverty for cities. Urbanization has led to better health and higher levels of education in many nations.[16] Countries that rank highest in freedom and human development are among the most urbanized.[17] These facts might suggest that rural villages house the world's worst suffering. But people often endure equal hardships in slums. Many poor citizens fear for their lives even as they live near the gated homes and armored cars of the wealthy.

The latest edition of *State of the World's Cities* reports on slum epidemics stoked by overcrowded housing and insufficient toilets.[18] Acute respiratory illness (primarily pneumonia) causes 18 percent of deaths of children under five and is easily spread in slums.[19] Surveys in Brazil, Peru, and Guatemala show more acute respiratory illness in slums than in rural areas.[20] Children in slums often play in human waste, which contains diarrhea-causing bacteria. Diarrhea causes a large share of the deaths of children in slums.[21]

In Nairobi, between 40 and 60 percent of the population squeezes into slums on 5 percent of the land, and people relieve themselves in bags that become "flying toilets" when tossed.[22] Some 27 percent of children in Nairobi slums have diarrhea, compared with 19 percent in rural Kenya.[23] In Khartoum, where slums are home to 80 percent of the residents, 33 percent of people in slums have diarrhea compared with 29 percent in rural Sudan.[24] In Bangladesh, one in four slum dwellers has diarrhea—twice the rural average.[25]

Rapid urbanization in sub-Saharan Africa has hastened HIV's spread as well. Many women resort to selling sex to earn money to feed their children; women are likelier than men to be stuck in slums with few employment prospects.[26] Surveys in Burkina Faso, Burundi, Ghana, Kenya, Mali, Niger, and Zambia show that HIV is twice as high in urban as in rural areas and that it is most prevalent in slums.[27]

Some governments use force to destroy slums. In May 2005, Zimbabwe mobilized its police and army in Operation Murambatsvina ("Drive Out Trash") to demolish slums.[28] UN Special Envoy Anna Tibaijuka of UN-HABITAT estimates that 700,000 people lost their homes or businesses in this action, which was popularly known as "Operation Tsunami"

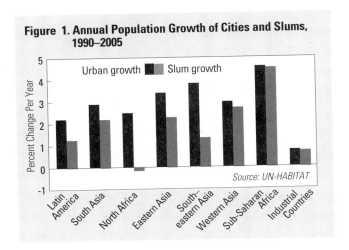

Figure 1. Annual Population Growth of Cities and Slums, 1990–2005

Source: UN-HABITAT

for its devastating speed.[29] Satellite images of Harare in June show empty spaces where similar shots in April showed structures.[30] In China, an official says the government moved more than 400,000 Beijing families into better housing between 1990 and 2003, in part to prepare for the 2008 Olympics; activists say 2.5 million people have been forcibly evicted since 1993 with little or no compensation.[31]

In 2000, as part of the Millennium Development Goals, world leaders pledged to improve the lives of 100 million slum dwellers—10 percent of the total—by 2020.[32] Of more than 100 countries surveyed by UN-HABITAT, only a handful are on track to meet this target, including Thailand, Egypt, Sri Lanka, and Tunisia.[33] Another 15, including Brazil, Mexico, South Africa, the Philippines, and Indonesia, have made limited progress.[34] UN-HABITAT finds that political commitment from top government leadership has spurred improvements in these cases.[35] But it is also clear that federations of slum dwellers and other nongovernmental organizations have pressured governments to respond.[36]

Action Needed on Water and Sanitation

Hilary French

As of 2002, the most recent year for which data are available, some 1.1 billion people worldwide lack access to an "improved water supply"—defined by the World Health Organization (WHO) and UNICEF as sources that are likely to provide safe drinking water, such as a household connection, a borehole, or a protected spring.[1] (See Figure 1.) And some 2.6 billion people are thought to lack access to "improved sanitation facilities," such as a connection to a public sewer, a septic system, or an improved pit latrine.[2] (See Figure 2.) Most people lacking access to both these basic necessities live in rural areas, but coverage is also scant in the burgeoning urban slums of the developing world.[3]

pp. 78, 110, 114

Taken together, this lack of access to clean drinking water and adequate sanitation poses a public health threat of staggering proportions. Half of all people in the developing world suffer from at least one of the main diseases linked to dirty water and inadequate sanitation—diarrhea, ascariasis, dracunculiasis (guinea worm disease), hookworm, schistosomiasis (also known as bilharzia or snail fever), and trachoma.[4] Diarrheal diseases linked to an unsafe water supply and inadequate sanitation and hygiene are by far the largest killer among these various scourges.[5] They alone claim some 1.6 million lives every year, most of them children.[6]

World leaders adopted a far-ranging set of Millennium Development Goals (MDGs) and targets at the U.N. Millennium Summit in 2000, including halving by 2015 the proportion of people without access to safe drinking water.[7] Two years later, at the World Summit on Sustainable Development in Johannesburg, governments agreed to add to the MDGs the goal of also halving by 2015 the share of the population lacking access to basic sanitation services.[8] At current rates of progress, analysts believe that the world is roughly on track to meet the global clean drinking water target by 2015—an encouraging accomplishment.[9] With sanitation, though, estimates suggest that the world is likely to miss the mark by a half-billion people.[10]

Although closing the water and sanitation gap is a formidable challenge, the picture is not uniformly grim. Significant progress was in fact made over the last decade, with more than 1 billion people added to the ranks of those with access to an improved water supply between 1990 and 2002—some 655 million of these people lived in urban areas and 436 million were in the countryside.[11] Overall, the share of people worldwide enjoying access to better water increased from 77 percent in 1990 to 83 percent in 2002.[12]

Improvement in this area occurred in all regions during the 1990s, but from different bases and at various rates.[13] Southern Asia, for example, registered the fastest progress, with the share of its population with access to improved water climbing from 71 percent in 1990 to 84 percent in 2002.[14] In Latin America and the Caribbean, this figure climbed from 83 to 89 percent in the same period, while in sub-Saharan Africa it rose from only 49 percent in 1990 to 58 percent in 2002.[15]

As a general matter, sub-Saharan Africa and the Pacific region have the lowest drinking water coverage rates, but in absolute terms Asia is home to the largest number of unserved people.[16] Although most regions are generally on schedule to meet the 2015 target, this is not the case in sub-Saharan Africa, where reaching this goal would require doubling the annual rate of increase in the number of people served in urban areas and tripling it in rural areas.[17]

Access to adequate sanitation services also expanded during the 1990s, with more than 1 billion people receiving coverage for the first time between 1990 and 2002—some 608 million of them in urban areas and 440 million in the countryside.[18] Overall, the share of people worldwide with access to improved sanitation facilities increased from 49 to 58 percent during this period.[19] Despite this progress, analysts project that if current trends continue, nearly 2.3 billion people will still lack access to adequate sanitation in 2015.[20] Reaching the MDG target would require increasing the population with access to improved sanitation by 138 million people annually between 2002 and 2015—a substantial increase over the current rate of progress.[21] And even if this can be done and the 2015 target met, an estimated 1.8 billion people will still lack access to decent sanitation services in part due to population growth.[22]

As with drinking water, there is considerable variation in sanitation coverage. Most regions registered some improvement during the 1990s, but, again, the rates of change differed.[23] The poorest coverage rates are found in sub-Saharan Africa, at 36 percent, and South Asia, at 37 percent.[24] In absolute terms, however, Asia faces the greatest challenge: more than half the people without access to improved sanitation live in China and India.[25]

Although the challenge of meeting the MDG water and sanitation targets is daunting, it is by no means insurmountable. In fact, many promising technologies and approaches are already in use around the world. Scaling up such innovations would yield large dividends in both human and financial terms.[26] Household water treatment systems, for instance, use a range of techniques to clean water on-site, including chlorination, solar disinfection, filtration, and the use of chemical disinfection powders.[27] Similarly, improved hygienic practices such as handwashing have the potential to reduce diarrheal diseases dramatically.[28]

Analysts estimate that achieving the water and sanitation targets would cost some $11.3 billion per year—a sizable sum, but one that would pay for itself many times over in health care savings, increased school attendance, time savings, enhanced worker productivity, and prolonged life spans.[29] According to WHO, the proposed $11.3-billion investment would generate an overall economic return of $84 billion.[30]

In an effort to accelerate progress, the United Nations General Assembly proclaimed

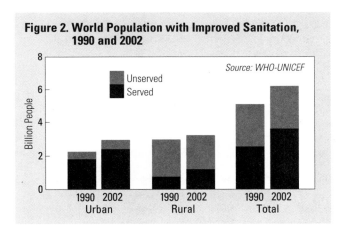

Figure 1. World Population with Improved Water Supply, 1990 and 2002

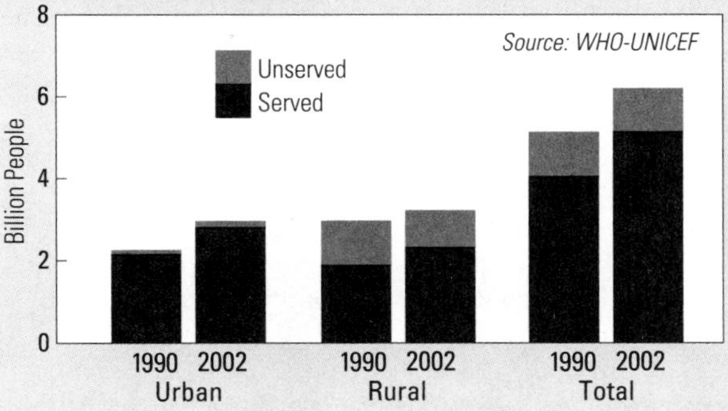

Source: WHO-UNICEF

Figure 2. World Population with Improved Sanitation, 1990 and 2002

Source: WHO-UNICEF

2005–15 the International Decade for Action: Water for Life.[31] Officially launched on March 22, 2005, the decade provides an umbrella for a wide range of ongoing water-related initiatives, including the 4th World Water Forum, which was held in Mexico City in March 2006.[32] If it serves its purpose, the International Decade for Action will jumpstart efforts by national governments, international institutions, civil society, the private sector, and others to put in place the innovative programs and policies needed to meet and—better yet—to exceed the 2015 water and sanitation targets.

Car-sharing Continues to Gain Momentum   *Susan A. Shaheen*

With auto ownership and fuel costs rising, people everywhere are seeking alternatives to private vehicle ownership. Car-sharing (or short-term vehicle rentals) provides such an alternative through hourly rates and subscription-access plans, especially for individuals and businesses in major cities with good access to other transportation modes, such as transit and carpooling.

LINKS pp. 42, 66

The principle of car-sharing is simple: individuals gain the benefits of private vehicle use without the costs and responsibilities of ownership. People involved in this typically join an organization that maintains a fleet of cars and light trucks in a network of locations, such as lots at transit stations or in neighborhoods or businesses.[1] Most car-sharing operators manage their services with some degree of modern computer-based technologies, which can include automated reservations, smart card vehicle access, and real-time vehicle tracking.[2]

For nearly 20 years, there has been growing worldwide participation in car-sharing. Some 330,000 individuals—nearly two thirds of whom are in Europe—now share at least 10,500 vehicles as part of organized car-sharing services.[3] (See Table 1 and Figure 1.) Many of these operations began in Switzerland and Germany in the late 1980s and later spread to 12 other countries on the continent and to the United Kingdom. In the 1990s, North America and Asia also started professional car-sharing activities. More recently, three car-sharing initiatives were launched in Australia starting in 2003.[4]

One of the first European initiatives can be traced to a cooperative known as "Sefage" (Selbstfahrergemeinschaft), which initiated services in Zurich, Switzerland, in 1948 and remained in operation until 1998.[5] This early effort was mainly motivated by economics. Individuals who could not afford to buy a car instead shared one. Elsewhere, a series of shared-car experiments were tried but later discontinued, such as Procotip in France from 1971 to 1973, Green Cars in the United Kingdom from 1977 to 1984, and programs in three Swedish cities at different times in the late 1970s through the 1990s.[6]

U.S. car-sharing began with two experiments: Mobility Enterprise, a Purdue University research program from 1983 to 1986, and the Short-Term Auto Rental (STAR) demonstration in San Francisco from 1983 to 1985.[7] More successful car-sharing operations worldwide began in Zurich in 1987 and in Berlin in 1988.[8]

A number of social and environmental benefits are commonly associated with car-sharing and are supported by an increasing body of empirical evidence, although differences in methodologies have produced inconsistent results. According to recent studies, sharing a car reduces the need for 4–10 privately owned cars in Europe and for 6–23 cars in North America.[9] North American studies and member surveys suggest that 11–29 percent of car-sharing participants sold a vehicle after joining a program, while 12–68 percent delayed or decided against buying a car.[10] Earlier European studies indicated a range of 16–32 percent of participants selling a vehicle after joining car-sharing; however, a more conservative range (23–26 percent), in contrast to North America, avoided or postponed a vehicle purchase.[11]

While the estimates of forgone vehicle purchases appear to be high in North America compared with Europe, it is important to note that they are based on "stated preference" survey responses, which can be overstated and are typically less reliable than "revealed preference" data (such as actual number of cars sold after joining car-sharing). Furthermore, auto ownership is much higher in the United States, so the potential to reduce the number of cars in

Table 1. Car-sharing Members and Vehicles, by Region, 2005

Region	Members	Vehicles
Europe	210,000	7,400
North America	105,571	2,409
Asia	13,500	726
Australia	600	35
Total	329,671	10,570

Source: Estimates based on discussions with regional experts and car-sharing organizations.

a household is presumably greater.[12]

European studies indicate a large reduction in vehicle-kilometers traveled (VKT), between 28 and 45 percent.[13] Data on VKT reduction range from as little as 7.6 percent to as much as 80 percent in Canada and the United States.[14] Estimates differ substantially between members who gave up vehicles after joining a program and those who gained initial access to a car through sharing one.[15] There was an average reduction of 44 percent in VKT per car-sharing user across North American studies.[16]

In Europe, car-sharing is estimated to reduce the average user's carbon dioxide emissions by 40–50 percent.[17] In addition, many car-sharing organizations include low- emission vehicles, such as gasoline-electric hybrid cars, in their fleets, which also reduces users' impacts on air quality and climate change.[18]

In addition, car-sharing shows evidence of beneficial social impacts. People can gain or maintain vehicle access without bearing the full costs of car ownership.[19] Depending on location and organization, the maximum annual mileage up to which car-sharing is more cost-effective than owning or leasing a personal vehicle lies between 10,000 and 16,093 kilometers.[20] Low-income households and college students also benefit from participating in car-sharing.[21]

Car-sharing continues to grow in business, transit, and fleet markets (such as government vehicles), as well as in university settings, particularly in North America. With few exceptions, advanced technology plays an important role in car-sharing worldwide. In Australia, France, the Netherlands, and North America, there is increased governmental interest and support for car-sharing, including supportive policies and grants. Competition among operators in the same region is also increasing, particularly in Germany and the United States.[22]

Future expansion in some regions may

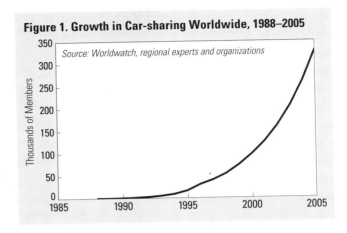

Figure 1. Growth in Car-sharing Worldwide, 1988–2005

Source: Worldwatch, regional experts and organizations

reflect a response to economic conditions, such as reduced household budgets and rising fuel prices. There will likely be entrants in new locations, such as Malaysia, South Africa, and New Zealand. Competition among operators is sure to continue, resulting in better services and choices for customers and, in some cases, in mergers and company closures. Along with competition, there will likely be increased cooperation among car-sharing operators and other partners, such as public transit (using smart card ticketing and access technologies, for instance), businesses, rental car companies, hotels, resorts, and shopping outlets.

Continued growth in business, fleet, transit, and university car-sharing markets is projected, as well as increased market share among households that need access to a second private vehicle. The neighborhood or individual car-sharing market is likely to grow as standards (on vehicle access technologies, for example) emerge that facilitate linkages or cross-agreements among operators in a region. In addition, car-sharing will be more widely integrated into urban transport and land use strategies through zoning variances for developers and supportive parking policies, for instance. And technological advances should encourage more people to join car-sharing programs that offer, for example, open-ended bookings, instant access, one-way rentals, satellite radio, and prepaid usage cards.[23]

Obesity Reaches Epidemic Levels

Peter Stair

Obesity—defined as a ratio of body mass to height squared totaling 30 or higher—has rapidly become a common disease in recent decades. Once a problem reserved for only the wealthiest, obesity now afflicts more than 300 million people.[1] The chronic ailments it can lead to—from cardiovascular disease to osteoporosis and depression—are a significant burden for many societies.

LINKS pp. 22, 24, 54, 64

Obesity first reached epidemic levels in wealthier northern countries, but it has since become pandemic across a range of diverse nations. In the United States, the prevalence more than doubled between 1990 and 2005, to about 40 percent.[2] China experienced a doubling too, between 1992 and 2002, though to a much lower prevalence rate of nearly 7 percent.[3] In Egypt and Kuwait, more than 30 percent of people are now obese.[4] Even among people in sub-Saharan Africa, where poverty is common, being overweight is a growing problem.[5]

Experts attribute this spread to a variety of factors, but basically the problem is that people are consuming more calories than they burn. Large-scale mechanized agriculture and economic growth have made energy-dense foods (meats, sugars, and oils) far more common. Global meat production per person has increased by more than 75 percent since 1961, while per capita sugar consumption has risen by more than 25 percent.[6] Over the same time, the price of a high-fat diet was cut in half.[7]

Food marketers have made processed foods—such as snacks, soda, and fast food—more available and alluring. The number of restaurants run by large fast-food chains has more than tripled since 1980, and soda consumption now rivals milk consumption globally.[8] "Empty calories" from concentrated sugar and fat fail to satisfy the appetite as effectively as high-fiber fruits and vegetables, so overeating is easier when such rich foods replace high-fiber meals.[9]

Mechanized agricultural systems have reduced the need for human labor on farms, while city life has offered an increasing amount of motorized transportation, including cars, elevators, and escalators.[10] Hard pavements and high crime rates can deter urban dwellers from outdoor activities, while sedentary pastimes such as watching television absorb at least three hours of the average person's day in the industrial world, more than half that individual's leisure time.[11]

In developing countries, city dwellers tend to be heavier. In China and Indonesia, the obesity rate is more than twice as high in cities as in rural areas. In Congo, the rate is six times higher in cities.[12] In already urbanized countries, however, where unhealthy, high-calorie foods can actually be cheaper and more accessible in poorer rural areas, city dwellers are often slimmer than their rural compatriots.[13] In 2003, people living in New York City were slightly less obese than the average American, and rural East Germans were more likely to be overweight than residents of Hamburg and Berlin.[14]

Certain populations also seem more likely to become overweight. Women generally have higher body mass index ratios than men, and in some countries, such as Kuwait, Russia, and Barbados, their obesity rate far exceeds that of men's—a disparity likely caused by their proximity to foods and lack of physical activity.[15] Some populations may also be more likely to have "thrifty genes" that store fat more readily.[16]

Probably because fat deposits were vital for surviving occasional periods of food scarcity in the Pacific Islands, almost half of all Samoans are obese today, as are more than 80 percent of Nauruans; in contrast, people from India or China tend to accumulate body mass less easily.[17] (See Figure 1.) Since these slimmer populations nonetheless manifest health complications at lower body mass index levels, public health officials have suggested different obesity thresholds for different ethnic groups.[18] Some argue that a waist-hip ratio, which is a measure of abdominal obesity, more accurately predicts the likelihood of health problems.[19]

Excess deposits of fat clog the body's metabolic system and strain the heart, bones, and ligaments. Being obese significantly raises a person's risk of cardiovascular disease, type II diabetes, respiratory diseases, osteoarthritis, gallbladder disease, and certain cancers (notably breast, colon, prostate, endometrium, kidney, and gallbladder cancers).[20] Obesity also reduces quality of life by contributing to sleep apnea, skin problems, infertility, loss of motivation, absenteeism, social isolation, and depression.[21] The global growth of type II diabetes has been particularly rapid—it has increased six times, to more than 177 million people, since 1985.[22]

In industrial countries, studies have found that as much as 12 percent of health care costs are attributable to obesity.[23] European Union studies have estimated the annual cost of obesity at about $40 billion (33 billion euros); if all problems of being overweight are counted, the figure is $155 billion (130 billion euros).[24] In the United States, the cost has been estimated at $75–118 billion, which rivals or exceeds the health-related costs of smoking.[25] A National Institute on Aging study warned that unless the disease is curtailed, obesity could contribute to the first sustained reduction in U.S. life expectancy during the modern era.[26] In developing countries, the rise in expenditures is adding strain to health care systems still trying to help those who are underweight.[27]

Weight loss products and programs now form a huge industry. In the United States alone, dieters spend $40–100 billion trying to lose weight, even though nearly all of them fail to lose pounds permanently.[28] Liposuction (surgical removal of fat cells) and gastric bypass (a reduction in the size of the stomach) are among the fastest-growing operations in the medical industry.[29]

Aiming to help children avoid early obesity, community health agents have focused on schools, requiring healthier lunches, the removal of vending machines, lessons on nutrition and cooking, and additional time for physical activities.[30] One successful campaign, Singapore's Fit and Trim program, combined nutrition classes, calorie-counting, exercise, and special counsel-

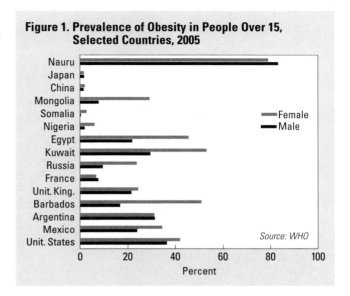

Figure 1. Prevalence of Obesity in People Over 15, Selected Countries, 2005

Source: WHO

ing for particularly overweight children, helping to reduce the childhood obesity rate by about 15 percent during the 1990s.[31]

Consumer groups have successfully pressured retailers such as McDonald's to reduce the availability of "super-sized" portions and to offer more salad and fruit options, but a U.S. Institute of Medicine report recently concluded the continued marketing of junk foods to children was a central cause of obesity.[32] Some European governments have already limited the marketing of unhealthy foods, while some academics have suggested direct taxation of nutritionally barren but high-calorie items.[33]

Others are promoting greater physical activity. Governments and political figures have launched publicity campaigns to encourage daily physical exercise and bicycle commuting.[34] Urban planners are beginning to design cities that encourage walking. Even corporations are getting involved: Sprint designed its company headquarters with slow elevators, to encourage employees to use the stairs, and placed the eating complex and parking lots at a good distance from the offices, in order to encourage daily walks.[35]

Corporate Responsibility Reports Take Root

Erik Assadourian

In 2004, nearly 1,800 transnational corporations (TNCs) or their affiliates filed reports on issues of corporate responsibility, up from virtually none in the early 1990s.[1] With some 1,600 reports already filed for 2005—estimated at about 85–90 percent of the likely total—this trend is on track to grow.[2] (See Figure 1.) These responsibility reports, sometimes referred to as nonfinancial reports, cover everything from labor standards and impacts on local communities to toxic releases and greenhouse gas emissions.

Thus far, most responsibility reports are filed by European corporations. Of those produced between 2001 and 2005, 54 percent came from Europe, 25 percent from Asia and Australia, 17 percent from North America, 2 percent from South America, and 2 percent from Africa and the Middle East.[3] Companies in developing economies, even major ones such as India and China, are producing few such reports. In 2004, only 5 Indian and 11 Chinese companies submitted them.[4]

Many of the largest TNCs now file annual nonfinancial reports. According to a 2005 KPMG survey, 52 percent of the top 250 companies of the Fortune 500 filed such a report in 2005, up from 45 percent in 2002.[5] If responsibility sections in financial reports are included, the total increases to 64 percent.[6]

Of companies listed on the FTSE 100, 83 filed significant reports in 2005.[7] Only one of the other 17 companies did not file at least a partial report on some of its own activities or those of its subsidiaries.[8] Of the S&P 100, 39 issued responsibility reports in 2004.[9] And a 2005 survey of 160 annual reports of companies selected randomly from the S&P 500 and the S&P/TSX Composite Index found that 36 percent of these reports included sections on issues of responsibility.[10]

As encouraging as this sounds, if roughly 1,800 TNCs or their affiliates are filing responsibility reports, that still leaves 97.5 percent of the nearly 70,000 TNCs worldwide without such documents.[11] Moreover, many of the reports filed are below par—lacking in details, transparency, or inclusion of long-term goals.[12]

"Most companies fail to give any real insight into what they are reporting on and why they are doing so," as noted in Risk & Opportunity, the 2004 Survey of Corporate Sustainability Reporting.[13] In 2003, less than 40 percent of the reports received any sort of third-party verification.[14]

Still, there are a few leaders in this field. CorporateRegister.com categorized about a quarter of the 1,783 documents filed in 2004 as full sustainability reports—ones that highlighted companies' efforts on environmental, social, economic, and community issues.[15]

Corporate responsibility reporting can serve as a central tool in helping companies reveal environmental and social weaknesses and provide strategies to remedy them. As the Chairman of Royal Dutch Shell, Jeroen van der Veer, explains "we have seen how, if done honestly, reporting forces companies to publicly take stock of their environmental and social performance, to decide improvement priorities, and deliver through clear targets."[16] By reporting, corporations admittedly expose their operations to more public scrutiny, yet they also increase trust among stakeholders—as long as they are actually working toward stated goals and not just making empty, unverifiable claims.

Some companies are using these reports not only to declare immediate impacts but also to state long-term goals and their yearly progress toward achieving them. For example, in 1998 BP announced the goal of cutting its greenhouse gas emissions to 10 percent below 1990 levels by 2010 and started publishing information on its annual releases.[17] By 2001 BP had already reached this goal—and in the process the company saved $650 million.[18]

Starbucks, too, has used its annual reports to declare a commitment to reduce its environmental and social impact through the creation of a sustainable coffee supply. In 2004, 19.7 million kilograms of its coffee (14.5 percent) were produced under its rigorous Coffee and Farmer Equity (C.A.F.E.) standards, up from 6 million kilograms the year before.[19] These standards, verified by an external auditor, award points for 28 key sustainability indicators, such as the amount of water, energy, and pesticides used and how equitably the profits are distributed among workers.[20] Starbucks' goal is to increase the share of C.A.F.E. standard coffee to 60 percent by 2007.[21]

Toyota Motor Corporation may be one of the leading companies in using reports to demonstrate both successful improvements and long-term commitment. For example, the company's 2005 report details reductions in discharges of waste, carbon dioxide, and toxic chemicals, along with increases in average fuel efficiency and sales of hybrid cars.[22] In addition, the report makes public the Fourth Toyota Environmental Action Plan for the period 2006 to 2010.[23] This comprehensive plan lists goals for a range of issues—from energy and climate change to waste, recycling, and broader engagement with society, such as helping to secure a broader public commitment to a "recycling-based society."[24]

One challenge with nonfinancial reporting is standardization. Currently no standard format has been established, so companies are often burdened by competing information demands from dozens of nongovernmental organizations. One leading guideline, the Global Reporting Initiative (GRI), has been adopted at least in name from more than 650 corporations (though some critics suggest the number of corporations that have truly incorporated GRI standards into their reports is closer to 300).[25] Established in 1997, the GRI has been regularly evolving, with new standards released in 2002 and in 2006.[26] Improved guidelines that incorporate feedback from hundreds of stakeholders will be released in mid- to late-2006.[27] The new version plans to include clearer, more comparable indicators and added instructions on how to prepare responsibility reports and better engage stakeholders.[28]

While growth in reporting continues, the pace is starting to slow.[29] One way to accelerate the growth will be to mandate corporate res-

Figure 1. Corporate Social and Environmental Reports, 1992–2005

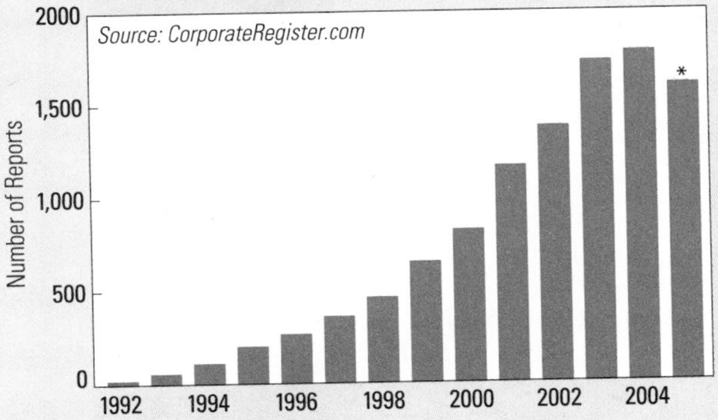

Source: CorporateRegister.com

* Data still being collected. This number represents 85–90 percent of the expected total.

ponsibility reporting. Already several countries have some reporting requirements. Denmark, with its Green Accounting Law in 1995, was the first in Europe to require environmental reporting for companies that have significant impact on the environment.[30] Similar legislation has been enacted in the Netherlands, Norway, and Sweden.[31]

Yet few countries have created comprehensive environmental and social reporting laws for all publicly traded companies listed on their national stock exchanges. In 2001, France became the first when it passed the Nouvelles Régulations Économiques.[32] Among other requirements, this obligates companies to report "on how the company takes into account the social and environmental consequences of its activities."[33]

The United Kingdom enacted similar legislation in March 2005.[34] This law required 1,300 companies to disclose environmental and social issues "as necessary."[35] But it was terminated eight months later by the Chancellor of the Exchequer, who argued that it would be too burdensome for companies to implement.[36] This will almost certainly slow adoption of nonfinancial reporting in the United Kingdom.

Nanotechnology Takes Off *Hope Shand and Kathy Jo Wetter*

LINKS p. 52

Nanotechnology—the manipulation of matter at the scale of atoms and molecules—has been called the transformational technology of the twenty-first century.[1] Experts predict that nanotech will revolutionize manufacturing across all industry sectors. Globally, billions of dollars are pouring into basic research; the number of nanotech-related scientific articles, patents, and investment portfolios is surging. Nanotech enthusiasts promise the greatest and greenest industrial revolution ever.

Worldwide, industry and governments now spend over $10 billion on nanotech R&D annually.[2] The National Science Foundation predicts that nanoscale technologies will capture a $1-trillion market by 2011.[3] Including funds requested for 2006, the U.S. government's National Nanotechnology Initiative has spent more than $5 billion on nanotech R&D since 2001 (see Table 1), the biggest publicly funded science endeavor since the Apollo moon shot.[4]

Nanotechnology is not a discreet industry sector but a range of techniques used to manipulate matter at the scale where size is measured in billionths of meters. A nanometer (nm) equals one billionth of a meter. It takes 10 atoms of hydrogen side-by-side to equal one nanometer. A DNA molecule is about 2.5 nm wide. A red blood cell is vast in comparison: about 5,000 nm in diameter. And a human hair is about 80,000 nm thick.

The real power of nanoscale science is the potential to merge disparate technologies that can operate at this scale, including biotechnology, cognitive sciences, informatics, and robotics. Enabled by nanotechnology, technological convergence will be the strategic platform for manufacturing, medicine, defense, agriculture, energy production, transportation, and communication.[5]

Nanotech's "raw materials" are the chemical elements of the Periodic Table—the building blocks of everything, both living and non-living. At the nanoscale, where quantum physics rules, substances exhibit new properties, like extraordinary strength, color changes, increased chemical reactivity, or electrical conductivity—characteristics they do not exhibit at larger scales. Nanoscale copper, for instance, becomes a highly elastic metal at room temperature—stretching to 50 times its original length without breaking.[6] Zinc oxide, usually white and opaque, is transparent at the nanoscale, and nano-aluminum can combust spontaneously.[7]

Companies are now manufacturing nanoparticles (chemical elements or compounds smaller than 100 nanometers) for use in hundreds of commercial products—from crack-resistant paints and stain-resistant pants to food additives, odor-eating socks, and transparent sunscreens. These products came to market in the absence of public debate and regulatory oversight. No government has developed a regulatory regime that addresses the nanoscale and its societal impacts.

Hundreds of nano-products are commercially available already, and that is only the beginning.[8] Nanotech offers the potential to develop stronger, lighter materials, low-cost solar cells, invisible sensors, faster computers with more memory, filters for cleaning contaminated water, cancer-killing molecules, and much more.

These small wonders will have colossal impacts, but not all of them will be welcome. While the technology's boosters praise the potential benefits, they have yet to confront the possible downsides. A growing number of scientific studies and government reports warn that engineered nanoparticles could pose unique risks to human health and the environment, yet few toxicological studies have been done.[9]

When reduced to the nanoscale, particles have a larger surface area that can make them more chemically reactive.[10] As particle size decreases and reactivity increases, a substance that may be inert at the micro or macroscale can assume hazardous characteristics at the nanoscale. One concern is that the increased reactivity could harm living tissue by giving rise to free radicals, perhaps causing inflammation, tissue damage, or tumor growth. Invisible and invasive, nanoparticles can slip past guardians of the body's immune system, across protective membranes such as the skin and the blood-brain barrier.

Nanobiotechnology, the merging of living

and non-living matter at the nanoscale, harnesses the self-replicating and self-organizing properties of living organisms for industrial uses. "Biological machines" may one day be released in the environment to sweep up carbon or produce food.[11] But the potential hazards related to control or misuse of synthetic life forms have not been addressed.

Researchers envision other nanobiotech products—such as implants and prostheses—to enhance human cognitive and physical performance. The goal of enhancing human performance raises troubling ethical issues, and disability rights activists warn of a looming gap between the technologically "improved" and "unimproved."[12]

Some nanotech proponents maintain that the potential benefits are too great to put on hold until scientific certainty on toxicity is achieved and ethical issues are addressed. For instance, nanotech could help meet the developing world's most pressing needs—cheap energy and clean water. Yet new designer materials also have the potential to topple commodity markets, disrupt trade, and eliminate jobs.[13] Commodity obsolescence could displace workers in the developing world who cannot respond to sudden demands for new skills or different raw materials. Most will not have access to job retraining programs.

National economies and workers who depend on primary export commodities will be out of luck and livelihoods, resulting in increased poverty and political instability—at least in the short term. For example, industry is designing nanoparticles to strengthen and extend the life of automobile tires and creating new nanomaterials that could substitute for natural rubber. Demand for natural rubber could plummet, devastating millions of rubber workers in Southeast Asia. The point is not that the status quo should be preserved—but that nanoscale technologies will bring huge socioeconomic disruptions for which society is not prepared.

In a just and judicious context, nanotech could be beneficial for the poor and the envi-

Table 1. Estimated Government R&D Investment in Nanotechnology, 2001–05

Region	2001	2002	2003	2004	2005
	(million dollars)				
European Union	~225	~400	~650	~950	~1,050
Japan	~465	~720	~800	~900	~950
United States	465	697	862	989	1,081
Others	~380	~550	~800	~900	~1,000
Total	~1,535	~2,367	~3,112	~3,739	~4,081

Source: The National Nanotechnology Initiative; investments are in current dollars.

ronment. But in a world where privatization of science and corporate concentration prevail, it is the pursuit of profits that is fueling the nanotech revolution—not human development needs. The world's largest transnational corporations, leading academic labs, and nanotech start-ups are racing to win monopoly control of tiny tech's colossal market. The grab for patents on nanoscale products and processes could mean mega-monopolies on the basic building blocks of the entire natural world. Researchers in many areas of the developing world may find access to the nanotech revolution blocked by patent tollbooths, obliging them to pay prohibitive licensing fees and royalties.[14] Thus if current trends continue, nanotech will widen the gap between rich and poor and further consolidate economic power in the hands of multinational corporations.

A wide debate is needed on the implications of nanoscale technologies and about democratic control of science and technology for the public good. When the root problems are poverty and social injustice, new technology is never the silver bullet. Now that the potential hazards of nanotech for environmental health and safety are beginning to come to public light, industry is begging to "eliminate regulatory uncertainty" by calling for voluntary regulations. But debate should not be confined to meetings of experts or focus solely on health and safety aspects. The broader social and ethical implications of nanotechnology need to be addressed.

Notes

GRAIN HARVEST FLAT (pages 22–23)

1. U.N. Food and Agriculture Organization (FAO), *FAOSTAT Statistical Database*, at apps.fao.org, updated 24 January 2006.
2. Ibid.
3. Ibid.
4. Share of cropland from FAO, op. cit. note 1; share of irrigated land is a Worldwatch calculation based on statistics for share of world's grain production that comes from irrigated land (60 percent) and ratio of yields on irrigated to nonirrigated land from "Overview," in Ruth S. Meinzen-Dick and Mark W. Rosegrant, eds., *Overcoming Water Scarcity and Quality Constraints* (Washington, DC: International Food Policy Research Institute, 2001).
5. Clive James, "Global Status of Commercialized Biotech/GM Crops: 2005 (Executive Summary)," *ISAAA Briefs*, No. 34 (Ithaca, NY: International Service for the Acquisition of Agri-Biotech Applications, 2005).
6. FAO, op. cit. note 1.
7. FAO, *Food Outlook* (Rome: December 2005), p. 2.
8. Ibid., pp. 6–7, 35.
9. Ibid.
10. Ibid.
11. Ibid., pp. 12, 34.
12. Ibid.
13. Ibid., pp. 3, 15–16; FAO, op. cit. note 1.
14. FAO, "Rice and Human Nutrition," *International Year of Rice 2004 Fact Sheet No. 3* (Rome: 2004).
15. FAO, op. cit. note 1; FAO, op. cit. note 7, p. 12; share of global production and trade from FAO, op. cit. note 1.
16. FAO, op. cit. note 7, p. 12.
17. Ibid., p. 10.
18. Ibid., pp. 6–8.
19. Ibid.
20. FAO, op. cit. note 1.
21. FAO, op. cit. note 7, p. 3.
22. FAO Newsroom, "Hunger Slows Progress Towards Millennium Development Goals: New FAO Report on World Hunger Urges Governments to Accelerate Hunger Reduction" (Rome: 22 November 2005); FAO, *The State of Food Insecurity in the World 2005* (Rome: 2005).
23. FAO, op. cit. note 7; FAO, op. cit. note 1.
24. FAO, op. cit. note 7, p. 2.
25. Ibid.

MEAT CONSUMPTION AND OUTPUT UP (pages 24–25)

1. U.N. Food and Agriculture Organization (FAO), *FAOSTAT Statistical Database*, at apps.fao.org; FAO, "Meat and Meat Products," *Food Outlook No. 3*, September 2005.
2. FAO, *FAOSTAT Statistical Database*, op. cit. note 1.
3. Christopher L. Delgado and Claire A. Narrod, *Impact of Changing Market Forces and Policies on Structural Change in the Livestock Industries of Selected Fast-Growing Developing Countries, Final Research Report of Phase I—Project on Livestock Industrialization, Trade, and Social-Health-Environmental Impacts on Developing Countries* (Rome: International Food Policy Research Institute and FAO, 2002).
4. FAO, *Food Outlook*, op. cit. note 1.
5. Ibid.
6. Ibid.
7. Ibid.
8. World Health Organization (WHO), "Cumulative Number of Confirmed Cases of Avian Influenza A/(H5N1) Reported to WHO," at www.who.int/csr/disease/avian_influenza, 11 April 2006.
9. FAO, *Food Outlook*, op. cit. note 1.
10. Ibid.
11. Ibid.
12. Ibid.
13. FAO, *FAOSTAT Statistical Database*, op. cit. note 1.
14. FAO, *Food Outlook*, op. cit. note 1.
15. Cees de Haan, Henning Steinfiedl, and Harvey Blackburn, *Livestock and the Environment: Finding a Balance*, report of a study coordinated by FAO, United States Agency for International Development, and World Bank (Brussels: European Commission Directorate-General for Development, 1997), p. 8.
16. The China-US Agro-Environmental Center of Excel-

lence, informational brochure (Beijing: 2003).

17. Nina Bonnelycke, Office of Water/Office of Waste-water Management/Water Permits Division, Animal Feeding Operation Program, U.S. Environmental Protection Agency, Washington, DC, discussion with author, February 2006.

18. Andrew Martin, "US Venture Hints at Brazil's Hog Farm Potential," *Chicago Tribune*, 14 June 2004.

19. WHO, "Bovine Spongiform Encephalopathy," Fact Sheet No. 113 (Geneva: 2002).

20. FAO and WHO from "Vietnam to Expand Restrictions to Fight Avian Flu," *CIDRAP News* (Center for Infectious Disease Research and Policy), 20 April 2005.

21. Ibid.

FISH HARVEST STABLE BUT THREATENED
(pages 26–27)

1. U.N. Food and Agriculture Organization (FAO), *FAOSTAT Statistical Database*, at apps.fao.org, updated 23 August 2005.

2. Ibid.; U.S Bureau of the Census, *International Data Base*, electronic database, Suitland, MD, updated 26 April 2005.

3. FAO, op. cit. note 1.

4. Ibid.

5. Ian Payne, *Fish and Biodiversity*, Biodiversity Brief 11, Biodiversity in Development Project, a joint initiative of IUCN–The World Conservation Union, the U.K. Department for International Development, and the European Commission, undated.

6. FAO, "Capture Production 1950–2003," *FISHSTAT*, at www.fao.org/fi/statist/fisoft/FISHPLUS.asp, updated March 2005.

7. Ibid.

8. FAO, op. cit. note 1.

9. Ibid.

10. Ibid.

11. Ibid.; Meryl Williams, *The Transition in the Contribution of Living Aquatic Resources to Food Security*, Food, Agriculture, and the Environment Discussion Paper 13 (Washington, DC: International Food Policy Research Institute, 1996).

12. FAO, op. cit. note 1.

13. Ibid.

14. Ransom A. Myers and Boris Worm, "Rapid Worldwide Depletion of Predatory Fish Communities," *Nature*, 15 May 2003, pp. 280–83; J. David Allan et al., "Overfishing of Inland Waters," *BioScience*, December 2005, pp. 1041–51.

15. Allan et al., op. cit. note 14.

16. Ibid.

17. "Shark Fin Industry Slams Hong Kong Boycott,"

Agence France-Presse, 2 December 2005.

18. Myers and Worm, op. cit. note 14; Juan Forero, "Hidden Cost of Shark Fin Soup: Its Source May Vanish," *Manta Journal*, 5 January 2006.

19. Callum M. Roberts et al., "Effects of Marine Reserves on Adjacent Fisheries," *Science*, 30 November 2001, pp. 1920–23.

20. Andrew Balmford et al., "The Worldwide Costs of Marine Protected Areas," *PNAS (Proceedings of the National Academy of Sciences)*, 29 June 2004, pp. 9694–97.

PESTICIDE TRADE SHOWS NEW MARKET TRENDS
(pages 28–29)

1. U.N. Food and Agriculture Organization (FAO), *FAOSTAT Statistical Database*, at apps.fao.org, updated 21 January 2006.

2. Ibid.

3. Ibid.

4. Ibid.

5. Ibid.

6. Ibid.

7. Ibid.

8. Ibid.

9. Ibid.

10. Ibid.

11. Ibid.

12. Ibid.

13. Ibid.

14. World pesticide estimate by David Pimental, Cornell University, Ithaca, NY, e-mail to author 6 February 2006.

15. Carina Weber and Susanne Smolka, *Towards Pesticide Use Reduction in Germany* (Hamburg, Germany: Pesticide Action Network Germany and Pesticide Action Network Europe, 2005).

16. World Health Organization in collaboration with U.N. Environment Programme, *Public Health Impact of Pesticide Use in Agriculture* (Geneva: 1990).

17. "Pesticides in Our Bodies," PAN North America, at www.panna.org/campaigns/bodyBurden.html, viewed 18 February 2006.

18. Pesticide Action Network Europe, *Pesticide Taxes – National Examples and Key Ingredients*, Briefing No. 6 (London: December 2005).

19. David Pimental, "Environmental and Economic Costs of the Application of Pesticides Primarily in the United States," *Environment, Development and Sustainability*, June 2005, pp. 229–52.

20. Stephanie Williamson, "Economic Costs of Pesticide Reliance," *Pesticides News*, September 2003, pp. 3–5.

21. Jennifer Curtis, *Fields of Change: A New Crop of Amer-*

ican *Farmers Finds Alternatives to Pesticides* (New York: Natural Resources Defense Council, 1998).

22. Leah C. M. Cuyno, George Norton, and Agnes Rola, "Economic Analysis of Environmental Benefits of Integrated Pest Management: A Philippine Case Study," *Agricultural Economics*, vol. 25, no. 2–3 (2001), pp. 227–33.

23. "Rotterdam Convention Enters into Force," *FAO Newsroom*, 24 February 2004.

24. "PIC COP-2 Elects Bureau, Completes Negotiations," *Linkages*, 30 September 2005.

25. Introduction and Overview of the Rotterdam Convention, at www.pic.int/en/print_it.asp?id=392, viewed 3 February 2006; "Ratifications, Signatories, and Conference Participants of the Convention," at www.pic.int/en/ViewPage.asp?id=265.

FOSSIL FUEL USE CONTINUES TO GROW (pages 32–33)

1. BP, *BP Statistical Review of World Energy* (London: 2005); International Energy Agency (IEA), *Oil Market Report*, 17 January 2006.

2. IEA, op. cit. note 1.

3. Ibid., p. 48.

4. Ibid.

5. Christopher Flavin and Gary Gardner, "China, India, and the New World Order," in Worldwatch Institute, *State of the World 2006* (New York: W. W. Norton & Company, 2006), p. 8.

6. U.S. Department of Energy (DOE), Energy Information Administration (EIA), *Short-Term Energy Outlook*, 10 January 2006; author's calculations based on DOE, EIA, *Monthly Petroleum Data*, Short-Term Energy Outlook Query System database, at tonto.eia.doe.gov/STEO_Query/app/papage.htm, 10 January 2006.

7. WRTG Economics, at www.wtrg.com/daily/crude oilprice.html; Klaus Matthies, "Commodity Prices at Record Level," *Intereconomics* (Hamburg Institute of International Economics), March/April 2005; DOE, EIA, "Imported Crude Oil Prices: Nominal and Real," *Short-Term Energy Outlook*, 7 March 2006.

8. Suzanne C. Hunt and Janet L. Sawin, with Peter Stair, "Cultivating Renewable Alternatives to Oil," in Worldwatch Institute, op. cit. note 5, p. 73.

9. Zijun Li, "China's Renewable Energy Law Takes Effect; Pricing and Fee-Sharing Rules Issued," *China Watch* (Washington, DC: Worldwatch Institute, 18 January 2006).

10. DOE, EIA, *Short-Term Energy Outlook*, op. cit. note 6; "OPEC Hopes for More Spare Capacity," *Daily Times*, 2 February 2006.

11. IEA, op. cit. note 1.

12. Ibid.

13. Ibid.

14. BP, op. cit. note 1.

15. Ibid.; Douglas Ogden, "We Don't Need More Power," *Newsweek International*, 6 February 2006.

16. BP, op. cit. note 1.

17. Ibid.

18. Author's calculations based on BP, op. cit. note 1; Eric Martinot, *Renewables 2005 Global Status Report* (Washington, DC: Worldwatch Institute, 2005).

19. "Gas Crisis Resolved but Lack of EU Energy Policy Remains Problem," *EURACTIV*, 4 January 2006.

20. "Ukraine Gas Row Hits EU Supplies," *BBC News*, 1 January 2006.

21. "Behind Rising Oil Cost: Nigeria," *Christian Science Monitor*, 19 January 2006.

22. Lawrence Kumins and Robert Bamberger, *Oil and Gas Disruption from Hurricanes Katrina and Rita* (Washington, DC: Congressional Research Service, 21 October 2005).

23. "Goldman Sachs: Oil Prices to Stay High for Years. Investment Bank Sees Crude Entering a 'Super Spike' Phase," *MSNBC*, 13 December 2005.

24. Crispin Aubrey, "The Emergence of the Post-Oil Era," *Wind Directions* (European Wind Energy Association), September/October 2005, p. 20.

25. Energy Sector Management Assistance Programme, *The Impact of Higher Oil Prices on Low Income Countries and on the Poor* (Washington, DC: World Bank, 2005).

26. Zijun Li, "The Cost of China's Energy Boom: Miners' Lives," *China Watch* (Washington, DC: Worldwatch Institute, 10 November 2005).

NUCLEAR POWER INCHES UP (pages 34–35)

1. Installed nuclear energy capacity is defined as reactors connected to the grid as of 31 December 2005 and is based on Worldwatch Institute database complied from statistics from the International Atomic Energy Agency (IAEA) and press reports primarily from *Associated Press*, *Reuters*, and *World Nuclear Association (WNA) News Briefing*, and from Web sites.

2. Worldwatch Institute database, op. cit. note 1.

3. International Energy Agency, *World Energy Outlook 2005* (Paris: 2005), p. 85.

4. Mycle Schneider and Antony Froggatt, "On the Way Out," *Nuclear Engineering International*, June 2005.

5. Worldwatch Institute database, op. cit. note 1.

6. Ibid.

7. "Germany: The Obrigheim Nuclear Power Plant Was Shut Down on 11 May," *WNA News Briefing*,

17 May 2005; "Sweden: The Barseback-2 Nuclear Power Reactor Was Permanently Shut Down," *WNA News Briefing*, 1 June 2005.

8. Worldwatch Institute database, op. cit. note 1.

9. "USA: The House of Representatives and the Senate Passed the Long-Awaited," *WNA News Briefing*, 2 August 2005.

10. "Canada: Ontario Power Generation's (OPG's) Refurbished Pickering A-1," *WNA News Briefing*, 4 October 2005.

11. "OPG Won't Fix Two Pickering Candus," *Electricity Daily*, 16 August 2005.

12. "Finland: Framatone ANP Hopes to Have a Revised Timetable for Construction," *WNA News Briefing*, 29 November 2005.

13. "Opening of Debate on New French Nuclear Reactor Cancelled," *Platts*, 17 October 2005.

14. "UK: The Government Will Make a Definitive Decision by the End of 2006," *WNA News Briefing*, 4 October 2005.

15. "Germany: On 11 November, Negotiators for the Country's Two Biggest Political Parties," *WNA News Briefing*, 22 November 2005.

16. "Spain: The National Congress Has Voted to Establish a Parliamentary Roundtable," *WNA News Briefing*, 24 May 2005.

17. Vladimir Slivyak, co-chairman, ECODEFENSE! and director, Anti-Nuclear Campaign, Socio-Economic Union, Nizhegorodskaya, Russia, e-mail to author, 17 January 2006.

18. Mika Ohbayashi, Director, Institute for Sustainable Energy Policies, Tokyo, e-mail to author, 26 January 2006.

19. "China to Build 31 Nuclear Plants Before 2020," *Power Engineering*, 21 September 2005.

20. Mohamed ElBaradei, "Nuclear Power: Preparing for the Future," *Statements of the Director General IAEA*, 21 March 2005.

21. "North Korea: The Korean Peninsula Energy Development Organization," *WNA News Briefing*, 29 November 2005.

22. "U.N. Security Council Gets Report on Iran," CNN.com, 7 February 2006.

WIND POWER BLOWING STRONG (pages 36–37)

1. Worldwatch estimate based on data from European Wind Energy Association (EWEA), "Wind Power Installed in Europe by End of 2005 (Cumulative)," at www.ewea.org, from American Wind Energy Association (AWEA), "U.S. Wind Industry Ends Most Productive Year," in *Wind Energy Weekly*, 27 January 2006, from "Canada's Wind Power Industry Shatters Growth Records in 2005," RenewableEnergyAccess.com, 15 February 2006, from Global Wind Energy Council (GWEC), "Record Year for Wind Energy: Global Wind Power Market Increased by 43% in 2005," press release (Brussels: 17 February 2005), from Birger Madsen, BTM Consult, Ringkøbing, Denmark, e-mail to author, 15 February 2006, and from BTM Consult, *World Market Update 2004* (Ringkøbing, Denmark: 2005).

2. Worldwatch estimate with data from EWEA, op. cit. note 1, from Madsen, op. cit. note 1, and from AWEA, "Wind Power: U.S. Installed Capacity (megawatts) 1981–2005," at www.awea.org.

3. Worldwatch estimates based on EWEA, op. cit. note 1, on AWEA, op. cit. note 1, on GWEC, op. cit. note 1, on Madsen, op. cit. note 1, and on BTM Consult, op. cit. note 1.

4. AWEA, op. cit. note 1; EWEA, op. cit. note 1; GWEC, op. cit. note 1; BTM Consult, op. cit. note 1.

5. AWEA, op. cit. note 1.

6. Ibid.

7. EWEA, op. cit. note 1; EWEA, "European Record for Wind Power: Over 6,000 MW Installed in 2005. Wind Energy Has Surpassed EC White Paper Targets for 2010," press release (Brussels: 1 February 2006).

8. Based on 2004 consumption levels; EWEA, op. cit. note 7.

9. EWEA, op. cit. note 1.

10. Based on 2001 consumption levels; Deutsches Windenergie-Institut GmbH, data for 2005, at www.dewi.de.

11. EWEA, op. cit. note 1.

12. EWEA, "Spain: Global Wind Power Leader Demonstrates How Wind Has Become a Mainstream Energy Source," press release (Brussels: 18 August 2005).

13. Ibid.

14. EWEA, op. cit. note 1.

15. Ibid.

16. Ibid.

17. Share of new capacity is Worldwatch estimate based on EWEA, op. cit. note 1, on AWEA, op. cit. note 1, on GWEC, op. cit. note 1, on Madsen, op. cit. note 1, and on BTM Consult, op. cit. note 1.

18. GWEC, op. cit. note 1.

19. Ibid.; Madsen, op. cit. note 1.

20. "China's Renewable Energy Potential: China Could Become World's Biggest Wind Energy Market by 2020," RenewableEnergyAccess.com, 8 November 2005.

21. Godfrey Chua, "Wind Power 2005 in Review, Outlook for 2006 and Beyond," RenewableEnergyAccess.com, 6 January 2006.

22. Ibid.; "FPL Energy, Iberdrola Deadlocked in Global Wind Power Lead," RenewableEnergyAccess.com, 24 February 2006.

23. BTM Consult, "Ten Year Review of the International Wind Power Industry 1995 to 2004; New Forecast for 2015; Long-Term Scenario for 2025," press release (Ringkøbing, Denmark: 19 October 2005) (this assumes that oil prices remain at $40–50 per barrel); "Wind to Reach 1 Million MW Capacity Within Two Decades, Report Predicts," ReFocus, 23 November 2005.

24. Madsen, op. cit. note 1.

25. "US/Canada Wind Power Markets and Strategies 2005–2010," Emerging Energy Research, December 2005, cited in Chua, op. cit. note 21.

26. "China's Renewable Energy Potential," op. cit. note 20.

SOLAR INDUSTRY STAYS HOT (pages 38–39)

1. Paul Maycock, PV News, various editions; Prometheus Institute, PV News, March 2006, pp. 4–5.

2. Worldwatch calculations based on Maycock, op. cit. note 1, and on Prometheus Institute, op. cit. note 1.

3. Prometheus Institute, op. cit. note 1.

4. Ibid.; Maycock, op. cit. note 1.

5. Marketbuzz 2006, cited in "Solarbuzz Reports World Solar Photovoltaic Market Grew 34% in 2005," Solarbuzz, 15 March 2005; exports calculated by Worldwatch with data from Prometheus Institute, op. cit. note 1, and from Marketbuzz 2006, op. cit. this note.

6. Arnulf Jäger-Waldau, "PV Status Report 2005: Research, Solar Cell Production and Market Implementation of Photovoltaics," produced for the European Commission, August 2005, p. 17.

7. Marketbuzz 2006, op. cit. note 5.

8. Prometheus Institute, op. cit. note 1, p. 5.

9. Ibid.

10. Ibid., pp. 4–5.

11. Ibid.

12. Calculated by Worldwatch with data from Paul Maycock, PV Energy Systems, e-mails to author, 27 January 2005 and 27 February 2006, from Maycock, op. cit. note 1, and from Prometheus Institute, op. cit. note 1.

13. Jesse Broehl, "In Wake of California Solar Plan, Industry Prepares for Expansion," RenewableEnergyAccess.com, 13 January 2006.

14. Ibid.; "Solar Power Shares Gain Ahead of California Energy Vote," Reuters, 12 January 2006.

15. Prometheus Institute, op. cit. note 1.

16. China from Maycock, op. cit. note 12, 27 February 2006; Suntech from Prometheus Institute, op. cit. note 1, p. 4.

17. Yingling Liu, "Shanghai Embarks on 100,000 Solar Roofs Initiative," China Watch (Worldwatch Institute), 10 November 2005.

18. Shortage from Paul Maycock, column in Prometheus Institute, op. cit. note 1, p. 8; more than 90 percent from Jesse W. Pichel and Ming Yang, "2005 Solar Year-end Review & 2006 Solar Industry Forecast," RenewableEnergyAccess.com, 11 January 2006.

19. Pichel and Yang, op. cit. note 18.

20. Shortages in 2006 from Marketbuzz 2006, op. cit. note 5.

21. By 2007 from Scott Sklar, "How Did the Silicon Shortage Situation Catch So Many by Surprise?" RenewableEnergyAccess.com, 9 February 2006; by 2008 from Catherine Lacoursiere, "Silicon Shortage Drives Global Solar Mergers & Acquisitions," RenewableEnergyAccess.com, 23 February 2006.

22. REN21 Renewable Energy Policy Network, Renewables 2005 Global Status Report (Washington, DC: Worldwatch Institute, 2005), p. 4.

23. Werner Weiss, Irene Bergmann, and Gerhard Faninger, Solar Heat Worldwide: Markets and Contribution to the Energy Supply 2004 (Paris: International Energy Agency, Solar Heating and Cooling Programme, March 2006).

24. Based on 2004 numbers, from Weiss, Bergmann, and Faninger, op. cit. note 23.

25. Domestic industry from James T. Areddy, "Heat for the Tubs of China," Wall Street Journal, 31 March 2006; share calculated by Worldwatch with data from REN21 Renewable Energy Policy Network, op. cit. note 22, p. 13.

26. Calculated by Worldwatch with data from REN21 Renewable Energy Policy Network, op. cit. note 22, p. 13.

27. Ibid.

28. Bundesverband Solarwirtschaft, cited in "Germany Opts for Solar Heat," Earthscan, 10 March 2006.

29. Ibid.

30. Some 30–70 percent of each building's water demand is to be met with solar thermal, with the fraction depending on demand volume and solar resources; European Solar Thermal Industry Federation, "Spain Approves National Solar Thermal Obligation," press release (Brussels: 21 March 2006).

BIOFUELS HIT A GUSHER (pages 40–41)

1. Christoph Berg, senior analyst, F.O. Licht, Agra Informa Ltd, Kent, U.K., e-mail to Peter Stair, World-

watch Institute, 25 January 2006. For comparison purposes, there are 159 liters per barrel. A ton contains 1,262 liters of ethanol or 1,136 liters of biodiesel. A liter of ethanol contains roughly 68 percent as much energy as a liter of gasoline, while a liter of biodiesel contains roughly 87 percent as much as a liter of diesel.

2. Ibid.
3. Ibid.
4. Production numbers from ibid.; gasoline consumption U.S. Department of Energy, Energy Information Administration, *Petroleum Information Monthly*, February 2006.
5. Berg, op. cit. note 1.
6. Ibid.
7. Ibid.
8. Ibid.
9. Ibid.; National Biodiesel Board, "Biodiesel Production Soars," press release (Jefferson City, MO: 8 November 2005).
10. "Biofuels and the International Development Agenda," *F.O. Licht's World Ethanol & Biofuels Report*, 8 July 2005.
11. Ibid.
12. Ibid.
13. Organisation for Economic Co-operation and Development (OECD), Working Party on Agricultural Policies and Markets, *Agricultural Market Impacts of Future Growth in the Production of Biofuels* (Paris: February 2006).
14. "Rising Ethanol Prices May Hit Brazilian Flex-fuel Car Sales," *F.O. Licht's World Ethanol & Biofuels Report*, 9 March 2006.
15. Jon Birger, "Wanna Make a Bet on Biofuels?" *Fortune*, 31 January 2006.
16. "World Ethanol Production 2005 to be Higher than Expected," *F.O. Licht's World Ethanol & Biofuels Report*, 24 October 2005.
17. Ibid.
18. Eric Martinot et al., *Renewables 2005 Global Status Report* (Washington, DC: Worldwatch Institute, 2005), p. 10.
19. BP, *Statistical Review of World Energy* (London: 2005).
20. "Biofuels and the International Development Agenda," op. cit. note 10.
21. Martinot et al., op. cit. note 18.
22. Chris Kraul, "Sweet Source of Growth," *Los Angeles Times*, 10 April 2006.
23. "Food Industry Calls for Biodiesel Alternatives," *Reuters*, 4 April 2006; Martin von Lampe, *Agricultural Market Impacts of Future Growth in the Production of Biofuels* (Paris: OECD, 2006), p. 15.

24. "USDA Raises Corn Use Estimate for Fuel Ethanol Production," *F.O. Licht's World Ethanol & Biofuels Report*, 8 March 2006.
25. Worldwatch Institute, "Biofuels for Transportation: Global Potential and Implications for Sustainable Agriculture and Energy in the 21st Century," prepared for the German Ministry of Food, Agriculture and Consumer Protection in coordination with the German Agency for Technical Cooperation and the German Agency of Renewable Resources (Washington, DC: 2006).
26. "Boom in US Fuel Ethanol Industry Shows No Sign of Slowing Down," *F.O. Licht's World Ethanol & Biofuels Report*, 21 February 2006.

CLIMATE CHANGE IMPACTS RISE (pages 42–43)

1. Goddard Institute for Space Studies, NASA, "Global Temperature Trends: 2005 Summation," January 2006, at data.giss.nasa.gov/gistemp/2005, viewed 15 March 2006.
2. Ibid.
3. Ibid.
4. NASA, "2005 Warmest Year in Over a Century," at www.nasa.gov/vision/earth/environment/2005_warmest.html, 24 January 2006, viewed 20 February 2006.
5. Data for 1996–2003 from Timothy Whorf, Scripps Institution of Oceanography, University of California, La Jolla, CA, e-mail to author, 18 January 2005; 2004–05 data from Dr. Stephen Piper, Carbon Dioxide Research Group, Scripps Institution of Oceanography, University of California, La Jolla, CA, e-mail to author, 13 March 2006.
6. Data for 1959–95 from C. D. Keeling and T. P. Whorf, "Atmospheric Carbon Dioxide Record from Mauna Loa," Carbon Dioxide Information Analysis Center, Scripps Institution of Oceanography, University of California, La Jolla, CA.
7. National Oceanic and Atmospheric Administration, "Climate of 2005—Annual Report," National Climatic Data Center, 13 January 2006, at www.ncdc.noaa.gov/oa/climate/research/2005/ann/global.html, viewed 21 March 2006.
8. Jill Hummels, "Greenland's Glaciers Moving Faster, Losing Mass," press release (Lawrence, KS: University of Kansas, 17 February 2006).
9. Paul R. Epstein and Evan Mills, *Climate Change Futures: Health, Economic and Ecological Dimensions* (Boston, MA: Harvard Medical School, Center for Health and Global Environment, 2005), p. 11.
10. Ibid.
11. Ibid., p. 101.

12. Stefan Lovgren, "Climate Change Creating Millions of 'Eco Refugees,' UN Warns," *National Geographic News,* 18 November 2005.

13. Doug Struck, "Inuit See Signs in Arctic Thaw," *Washington Post,* 22 March 2006.

14. Data for 2001–05 calculated by Worldwatch with data from BP, *BP Statistical Review of World Energy* (London: 2005).

15. Urs Siegenthaler et al., "Stable Carbon Cycle–Climate Relationship During the Late Pleistocene," *Science,* 25 November 2005, pp. 1313–17.

16. Elizabeth Kolbert, "Annals of Science: The Climate of Man," *The New Yorker,* three-part series, 25 April, 2 May, and 9 May 2005.

17. American Association for the Advancement of Science, "Carbon Dioxide Level Highest in 650,000 Years," press release (Washington, DC: 24 November 2005); "Climate Change: The Big Emitters," *BBC News,* 4 July 2005.

18. Natural Resources Defense Council, *Climate Change: In-Depth* (Washington, DC: 5 January 2005).

19. International Energy Agency, *World Energy Outlook 2004* (Paris: 2004).

20. David Ignatius, "Is It Warm in Here? We Could Be Ignoring the Biggest Story in Our History," *Washington Post,* 18 January 2006; "Global Warming: The Signs and the Science," *PBS Documentary,* 2 November 2005.

21. "FACTBOX: Europe's Emissions Trading Scheme," *Reuters,* 28 November 2005.

22. "Lessons Learned in 2005," *Point Carbon,* 24 January 2006.

23. Pew Center on Global Climate Change, *Learning from State Action on Climate Change* (Arlington, VA: March 2006).

WEATHER-RELATED DISASTERS AFFECT MILLIONS (pages 44–45)

1. Munich Reinsurance Company (Munich Re), NatCatSERVICE, e-mail to author, 19 January 2006.

2. Ibid.

3. Munich Re, *Topics Geo Annual Review: Natural Catastrophes 2005* (Munich: 2006), p. 7; Munich Re, *Hurricanes—More Intense, More Frequent, More Expensive* (Munich: 2006), p. 4.

4. Munich Re, *Hurricanes,* op. cit. note 3, p. 4.

5. Ibid., p. 7; Thomas Knutson, "Impact of CO_2-induced Warming on Simulated Hurricane Intensity to the Choice of Climate Model and Convective Parameterization," *Journal of Climate,* 15 September 2004, pp. 3477–95.

6. Munich Re, *Hurricanes,* op. cit. note 3, pp. 7–8;

C. D. Hoyos et al., "Deconvolution of the Factors Contributing to the Increase in Global Hurricane Intensity," *Science Express,* 16 March 2006.

7. Worldwatch calculation based on Center for Research on the Epidemiology of Disasters (CRED), *EM-DAT: The OFDA/CRED International Disaster Database,* viewed 9 March 2006. CRED data are continuously revised; the calculations include drought, extreme temperature events, floods, slides, wildfires, and windstorms.

8. CRED, op. cit. note 7. While fatality estimates vary between CRED and Munich Re, both show a decrease in fatalities from 2004 to 2005.

9. International Federation of Red Cross and Red Crescent Societies, *World Disasters Report 2005* (Geneva: 2005), p. 53.

10. Munich Re, *Topics Geo Annual Review,* op. cit. note 3, p. 24.

11. Hillary Rosner, "Rain Forest Gets Too Much Rain, and Animals Pay the Price," *New York Times,* 7 March 2006.

12. Ibid.

13. U.N. International Strategy for Disaster Reduction, "Governments Must Accelerate Their Efforts to Make Disaster Risk Reduction a National Priority," press release (Geneva: 18 January 2006).

14. Munich Re, *Topics Geo Annual Review,* op. cit. note 3, p. 32.

15. U.N. International Strategy for Disaster Reduction, op. cit. note 13.

16. Ibid.

17. International Federation of Red Cross and Red Crescent Societies, op. cit. note 9, p. 44.

18. Ibid.

19. Citizens' Disaster Response Center, "Programs and Services," at www.cdrc-phil.org.

HYDROPOWER REBOUNDS SLIGHTLY (pages 46–47)

1. BP, *BP Statistical Review of World Energy 2005* (London: 2005).

2. Ibid.

3. Eric Martinot, *Renewable 2005 Global Status Report* (Washington, DC: Worldwatch Institute, 2005).

4. BP, op. cit. note 1.

5. Ibid.

6. U.S. Department of Energy (DOE), Energy Information Administration (EIA), *International Energy Outlook 2005* (Washington, DC: 2005).

7. Martinot, op. cit. note 3.

8. BP, op. cit. note 1.

9. "Three Gorges Dam to be Completed by May," *China News,* 10 January 2006.

Notes

10. DOE, National Renewable Energy Laboratory, *Evaluation of China's Energy Strategy Options* (Golden, CO: 2005).
11. DOE, op. cit. note 6.
12. International Rivers Network (IRN), "World Bank 'New Investment Framework': A Great Leap Backwards for Sustainable Energy," press release (Berkeley, CA: 6 December 2005).
13. BP, op. cit. note 1; "Record Drought Dims Hydropower Outlook in US Northwest," *Reuters*, 17 May 2004.
14. DOE, EIA, "Renewable & Alternative Fuels, U.S. Data," *Electric Power Monthly*, February 2006
15. Duncan Graham-Rowe, "Hydroelectric Power's Dirty Secret Revealed," *New Scientist*, 26 February 2005.
16. Ibid.
17. Ibid.
18. World Commission on Dams, *Dams and Development: A New Framework for Decision-Making* (London: Earthscan, 2000), p. 278.
19. "Damming Evidence," *The Economist*, 17 July 2003.
20. IRN, op. cit. note 12.

ENERGY PRODUCTIVITY GAINS SLOW (pages 48–49)

1. International Monetary Fund (IMF), *World Economic Outlook Database*, September 2005; BP, *Statistical Review of World Energy* (London: 2005).
2. IMF, op. cit. note 1; BP, op. cit. note 1.
3. A. Grubler, *Technology and Global Change* (Cambridge, U.K.: Cambridge University Press, 1998).
4. IMF, op. cit. note 1; BP, op. cit. note 1.
5. BP, op. cit. note 1.
6. Tatyana P. Soubbotina, *Beyond Economic Growth*, 2nd ed. (Washington, DC: World Bank, 2004), p. 50.
7. H. Geller et al., "Polices for Increasing Energy Efficiency: Thirty Years of Experience in OECD Countries," *Energy Policy*, March 2006, pp. 556–73.
8. U.S. Environmental Protection Agency, Energy Star, "Compact Fluorescent Lightbulbs," at www.energystar.gov/index.cfm?c=cfls.pr_cfls.
9. J. W. Sun, "Three Types of Decline in Energy Intensity—An Explanation for the Decline of Energy Intensity in Some Developing Countries," *Energy Policy*, September 2003, pp. 519–26.
10. Lee Schipper, discussion with author, 30 March 2006.
11. BP, op. cit. note 1.
12. Geller et al., op. cit. note 7.
13. Ibid.
14. Ibid.
15. Frede Hvelplund and Henrik Lund, "Rebuilding without Restructuring the Energy System in East

Germany," *Energy Policy*, June 1998, pp. 535–46; U.S. Department of Energy, Energy Information Administration, *International Energy Outlook 2005* (Washington, DC: 2005).
16. Petra Opitz, "The (Pseudo-) Liberalisation of Russia's Power Sector: The Hidden Rationality of Transformation," *Energy Policy*, March 2000, pp. 147–55.
17. BP, op. cit. note 1.
18. Scott Murtishaw and Lee Schipper, "Disaggregated Analysis of US Energy Consumption in the 1990s: Evidence of the Effects of the Internet and Rapid Economic Growth," *Energy Policy*, July 2001 pp. 1335–56.
19. Geller et al., op. cit. note 7.
20. IMF, op. cit. note 1; BP, op. cit. note 1.
21. IMF, op. cit.note 1; BP, op. cit. note 1.
22. Jiang Lin, "Energy Conservation Investments: A Comparison between China and the US," *Energy Policy*, in press; 13 percent from IMF, op. cit. note 1, and from BP, op. cit. note 1.
23. IMF, op. cit. note 1; BP, op. cit. note 1.

GLOBAL ECONOMY GROWS AGAIN (pages 52–53)

1. International Monetary Fund (IMF), *World Economic Outlook Database* (Washington, DC: September 2005). Note the 2005 figure is a preliminary estimate from September 2005 and is subject to change.
2. IMF, op. cit. note 1.
3. Ibid.
4. Ibid.
5. Raghuram Rajan, "Transcript of the World Economic Outlook Press Briefing" (Washington, DC: IMF, 21 September 2005).
6. Ibid.
7. Ibid.
8. P. J. Webster et al., "Changes in Tropical Cyclone Number, Duration, and Intensity in a Warming Environment," *Science*, 16 September 2005, pp. 1844–46; C. D. Hoyos et al., "Deconvolution of the Factors Contributing to the Increase in Global Hurricane Intensity," *Science Express*, 16 March 2006.
9. IMF, op. cit. note 1.
10. Rajan, op. cit. note 5.
11. IMF, op. cit. note 1.
12. IMF, *World Economic Outlook 2005* (Washington, DC: 2005), p. 34.
13. IMF, op. cit. note 1; IMF, op. cit. note 12, p. 30.
14. IMF, op. cit. note 1.
15. IMF, op. cit. note 12, p. 9.
16. Population data from U.S. Bureau of the Census, *International Data Base*, electronic database,

Suitland, MD, updated August 2005; IMF, op. cit. note 1.

17. IMF, op. cit. note 1; IMF, op. cit. note 12, pp. 48–49.

18. IMF, op. cit. note 1; Census Bureau, op. cit. note 16.

19. IMF, op. cit. note 1; Census Bureau, op. cit. note 16.

20. IMF, op. cit. note 1; Census Bureau, op. cit. note 16.

21. IMF, op. cit. note 1; Census Bureau, op. cit. note 16.

22. U.N. Development Programme, *Human Development Report 2005* (New York: Oxford University Press, 2005), p. 270.

23. Ibid., p. 230.

24. Jason Venetoulis and Cliff Cobb, *The Genuine Progress Indicator 1950–2002 (2004 Update)* (Oakland, CA: Redefining Progress, 2004).

25. Ibid.

26. Ibid.

27. David Woodward and Andrew Simms, *Growth Isn't Working* (London: New Economics Foundation, 2006).

ADVERTISING SPENDING SETS ANOTHER RECORD (pages 54–55)

1. Robert Coen, "Insider's Report: Robert Coen Presentation on Advertising Expenditures," December 2005, at www.universalmccann, viewed 8 February 2006; recent data for Figure 1 from ibid.; historical data from Robert Coen, *Estimated World Advertising Expenditures*, as cited in Erik Assadourian, "Advertising Spending Stays Nearly Flat," in Worldwatch Institute, *Vital Signs 2003* (New York: W. W. Norton & Company, 2003), pp. 48–49.

2. Coen, "Insider's Report," op. cit. note 1.

3. Ibid.

4. Ibid.

5. ZenithOptimedia, "Ad Growth Stable with Healthy Hotspots," press release (London: December 2005).

6. Ibid. Note: the percentages in this paragraph use $394 billion as the total for major media ad spending, as $10 billion was in undifferentiated spending.

7. ZenithOptimedia, op. cit. note 5.

8. Ibid.

9. Ibid.

10. Ibid.

11. Ibid.

12. Ibid.

13. Ibid.

14. Coen, "Insider's Report," op. cit. note 1; population from U.S. Bureau of Census, *International Data Base*, electronic database, Suitland, MD, updated 26 April 2005.

15. Coen, "Insider's Report," op. cit. note 1; Census Bureau, op. cit. note 14.

16. Coen, "Insider's Report," op. cit. note 1; Census Bureau, op. cit. note 14.

17. Craig Endicott, "Global Marketing," *Advertising Age*, 14 November 2005, pp. 1–53.

18. Ibid.

19. Ibid.

20. Laurel Wentz, "Europe Stops School Soft-Drink Marketing," *AdAge.com*, 31 January 2006.

21. Ibid.

22. Ira Teinowitz, "Food Giants Targeted in $2 Billion Lawsuit," *AdAge.com*, 18 January 2006; Center for Science in the Public Interest (CSPI), "Parents and Advocates Will Sue Viacom & Kellogg," press release (Washington, DC: 18 January 2006); CSPI, "Letter of Intent to Sue," 18 January 2006, at cspinet.org/new/pdf/viacom___kellogg.pdf, viewed 10 February 2006.

23. Commercial Alert, "Prescription Drug Marketing Harms the Doctor-Patient Relationship," at stop drugads.org/learn_more.html, viewed 10 February 2006; Commercial Alert, "200+ Medical School Professors Call for End to DTC Prescription Drug Ads," press release (Portland, OR: 27 October 2005); Rich Thomaselli, "47% of Doctors Feel Pressured by DTC Drug Advertising," *Advertising Age*, 14 January 2003.

24. Henry J. Kaiser Family Foundation, *Impact of Direct-to-Consumer Advertising on Prescription Drug Spending* (Menlo Park, CA: 2003).

25. Commercial Alert, "200+ Medical School Professors," op. cit. note 23; Commercial Alert, "Prescription Drug Marketing," op. cit. note 23; Michelle Cottle, "Selling Shyness," *The New Republic*, 2 August 1999; Ray Moynihan and Alan Cassels, *Selling Sickness: How the World's Biggest Pharmaceutical Companies Are Turning Us All Into Patients* (New York: Nation Books, 2005), pp. 119–38.

STEEL OUTPUT UP BUT PRICE DROPS (pages 56–57)

1. International Iron and Steel Institute (IISI), "2005 (Full Year) Crude Steel Production," press release (Brussels: 18 January 2006).

2. Greg Mazurkiewicz, "Steel Prices Come Back Down to Earth," *Air Conditioning, Heating, Refrigeration News*, 17 November 2005; historical prices from CRUspi Global Index, CRU International, at www.cruspi.com.

3. J. Mehra, "Steel Industry: to Continue Shining in 2006," *The Hindu Business Line*, 17 January 2006; Bureau of Labor Statistics, U.S. Department of Labor, "U.S. Export Price Indexes and Percent Changes for Selected Categories of Goods: Decem-

Notes

ber 2004–December 2005," updated on 12 January 2006.

4. "China Still Leading Steel Demand Rise, But Oil Prices a Worry," *Engineering News*, 26 January 2006.

5. "Chinese Steel Output Surged 25% to a Record Last Year," *Bloomberg*, 19 January 2006.

6. IISI, op. cit. note 1.

7. "Chinese Steel Output Surged 25% to a Record Last Year," op. cit. note 5.

8. Steve Mackrell, "Steel, The 'New Textiles' for China," *Asian Times Online*, 5 November 2005.

9. "Forging a New Shape," *The Economist*, 8 December 2005.

10. "Statistics on China's Steel Imports and Exports in 2005," *Shanghai Securities News*, 10 January 2006.

11. "Forging a New Shape," op. cit. note 9.

12. Ibid.

13. "China Regroups Regional Steel Makers," *Xinhua News Agency*, 7 December 2005.

14. "Steel Glut Triggers Output Cuts," *BBC News*, 4 July 2005.

15. Ibid.

16. "Chinese Mills to Slash Production," *Steel Times International*, 24 November 2005.

17. IISI, op. cit. note 1.

18. Ibid.

19. "Steelmaking Raw Material and Input Costs," at www.steelonthenet.com.

20. Steve Mackrell, "Your Steel Trash, China's Treasure," *Asian Times Online,* 21 January 2006.

21. Ibid.

22. Ibid.

23. U.S. Geological Survey, *Mineral Commodity Summaries* (Washington, DC: January 2006).

24. Ibid.

ALUMINUM PRODUCTION INCREASES STEADILY (pages 58–59)

1. U.S. Geological Survey (USGS), "Aluminum," *Mineral Commodity Summaries 2006* (Washington, DC: Government Printing Office (GPO), 2006), pp. 22–23; Patricia Plunkert, aluminum specialist, USGS, Reston, VA, e-mail to author, 4 January 2006.

2. Patricia Plunkert, aluminum specialist, USGS, Reston, VA, e-mail to author, 3 December 2005.

3. USGS, op. cit. note 1; U.S. Bureau of the Census, *International Data Base*, electronic database, updated 26 April 2005.

4. USGS, "Bauxite and Aluminum," *Mineral Commodity Summaries 2005* (Washington, DC: GPO, 2005), pp. 30–31.

5. Wayne Wagner, "Aluminum," *Canadian Minerals Yearbook, 2003* (Ottawa, Canada: Natural Resources Canada, 2004), p. 9.11.

6. USGS, op. cit. note 4.

7. Ibid.

8. Wagner, op. cit. note 5.

9. Wayne Wagner, "Aluminum," *Canadian Minerals Yearbook, 2000* (Ottawa, Canada: Natural Resources Canada, 2001), p. 8.20.

10. USGS, *Mineral Commodity Summaries 2001–2006* (Washington, DC: GPO, various years).

11. Ibid.

12. European Aluminium Association, *Market Report 2004* (Brussels: 2005), p. 13.

13. Ibid., p. 7.

14. Patricia A. Plunkert, "Aluminum," *Engineering & Mining Journal*, April 2000, p. 39.

15. USGS, op. cit. note 1.

16. Aluminum Association, "In Depth Information/Recycling Process," at www.aluminum.org/Content/NavigationMenu/The_Industry/Recycling/In-depth_information/In-depth_information.htm, viewed 12 December 2005.

17. USGS, "Aluminum," *Minerals Yearbook—2004* (Washington, DC: GPO, 2005), p. 5.1.

18. Ibid.

19. International Aluminium Institute (IAI), "History of Aluminium," at www.world-aluminium.org/history/index.html, viewed 12 December 2005.

20. Ibid.

21. IAI, "Electrical Power Used in Primary Aluminium Production—Form ES002," at www.world-aluminium.org/iai/stats/index.asp, viewed 12 December 2005; the average U.S. household used 10,650 kilowatt-hours in 2001, according to U.S. Department of Energy (DOE), Energy Information Administration (EIA), *US Household Electricity Report* (Washington, DC: 2005).

22. Smelter energy use is a Worldwatch estimate, based on production data from USGS, op. cit. note 1, on electricity consumption per ton of aluminum production from IAI, op. cit. note 21, and on world electricity consumption data from DOE, EIA, "World Total Net Electricity Consumption, 1980–2003," at www.eia.doe.gov/emeu/iea/elec.html.

23. IAI, "Climate Change," at www.world-aluminium.org/environment/climate/index.html, viewed 12 December 2005.

24. IAI, op. cit. note 21.

25. Wayne Wagner, aluminum specialist, Minerals and Metals Sector, Natural Resources Canada, Ottawa, e-mail to author, 13 December 2005.

26. Container Recycling Institute, "Aluminum Can Recycling Rates (1990–2004)," at www.container-re

cycling.org/alumrate/CRIvsAA90-04.htm, viewed 12 December 2005; U.S. Environmental Protection Agency, *Characterization of Municipal Solid Waste in the United States: Source Data on the 2003 Update* (Washington, DC: 2000).

27. Energy savings is a Worldwatch estimate based on IAI, op. cit. note 21, on Aluminum Association, op. cit. note 16, and on DOE, op. cit. note 21.

ROUNDWOOD PRODUCTION HITS A NEW PEAK (pages 60–61)

1. U.N. Food and Agriculture Organization (FAO), *FAOSTAT Statistical Database*, at faostat.fao.org, updated 12 August 2005.
2. Ibid.
3. Ibid.
4. Ibid.
5. Ibid.
6. Ibid.
7. U.N. Economic Commission for Europe (UNECE)/ FAO, "Forest Products Annual Market Review, 2004–2005," *Timber Bulletin*, vol. LVII, no. 3 (2005).
8. FAO, op. cit. note 1.
9. J. L. Bowyer, "Changing Realities in Forest Sector Markets," *Unasylva*, vol. 55, no. 4 (2004).
10. FAO, op. cit. note 1.
11. Ibid.; Graeme Lang and Cathy Hiu Wan Chan, "China's Impact on Southeast Asian Forests," *CSR Asia Weekly*, 22 June 2005.
12. FAO, op. cit. note 1.
13. O. Hashiramoto, J. Castano, and S. Johnson, "Changing Global Picture of Trade in Wood Products," *Unasylva*, vol. 55, no. 4 (2004).
14. FAO, op. cit. note 1.
15. Ibid.
16. Ibid.
17. Ibid.
18. UNECE/FAO, op. cit. note 7.
19. Ibid.
20. Ibid.
21. Hashiramoto, Castano, and Johnson, op. cit. note 13.
22. Tachrir Fathoni, "Indonesian Experience: The Efforts in Protecting Forests and Forest Product," International Network for Environmental Compliance and Enforcement, Washington, DC, October 2004.
23. Hashiramoto, Castano, and Johnson, op. cit. note 13.

VEHICLE PRODUCTION CONTINUES TO EXPAND (pages 64–65)

1. Colin Couchman, Global Insight Automotive Group, e-mail to author, 7 February 2006.

2. Ibid.; Colin Couchman, e-mail to author, 13 January 2005; Global Insight Automotive Group, *Global Sales of Light Vehicles by Region & Country December 2003* (London: 2003); Ward's Communications, *World Motor Vehicle Data 2005* (Southfield, MI: 2005), p. 214; American Automobile Manufacturers Association, *World Motor Vehicle Facts and Figures 1998* (Washington, DC: 1998). Once intended for commercial purposes, light trucks have come to be used increasingly for passenger transport, particularly in the United States, although the use of such vehicles varies from country to country.
3. Ward's Communications, op. cit. note 2, p. 236. This source has fleet data for "passenger cars" and "commercial vehicles" but does not break out "light trucks" that are now typically used for passenger transportation. Because "commercial vehicles" includes vehicles suitable for goods but not passenger transport—although in relatively small numbers—including this category yields somewhat of an overcount. Yet limiting the count strictly to passenger cars would result in a more distorting undercount.
4. Calculated from Ward's Communications, op. cit. note 2, pp. 237–40.
5. Ibid.
6. Ward's Communications, *Ward's Motor Vehicle Facts & Figures 2005* (Southfield, MI: 2005), p. 14.
7. Ward's Communications, op. cit. note 2, p. 199.
8. Ibid., p. 17.
9. Ibid., p. 21.
10. Calculated by author from Ward's Communications, op. cit. note 6, p. 15.
11. Ibid.
12. Micheline Maynard, "A Comeback for the Car Species," *New York Times*, 5 January 2006.
13. Micheline Maynard, "Ford Eliminating up to 30,000 Jobs and 14 Factories," *New York Times*, 24 January 2006.
14. Micheline Maynard, "Toyota Shows Big Three How It's Done," *New York Times*, 13 January 2006.
15. "Toyota Says It Plans Eventually to Offer an All-Hybrid Fleet," *New York Times*, 14 September 2005.
16. "Sales Continue to Speed Up," *Business Week Online*, 28 December 2005; sales in 2005 are through November only.
17. "Toyota Says It Plans Eventually to Offer an All-Hybrid Fleet," op. cit. note 15.
18. Matthew L. Wald, "Designed to Save, Hybrids Burn Gas in Drive for Power," *New York Times*, 17 July 2005; Jeff Sabatini, "The Hybrid Emperor's New Clothes," *New York Times*, 31 July 2005.
19. Keith Bradsher, "China Preparing to Tax Vehicles

With Large Engines," *New York Times*, 26 August 2005.
20. Ibid.
21. U.S. Department of Energy, Energy Information Administration, *International Energy Annual 2003* (Washington, DC: 2005), Table 35.
22. Ward's Communications, op. cit. note 6, p. 84.
23. Danny Hakim and John M. Broder, "Slight Shift in S.U.V.'s in New Rule on Mileage," *New York Times*, 24 August 2005.
24. Ibid.

BICYCLE PRODUCTION UP (pages 66–67)

1. Global production numbers from United Nations, *The Growth of World Industry*, 1969 Edition, Vol. II (New York: 1971), from United Nations, *Yearbooks of Industrial Statistics*, 1979 and 1989 Editions, Vol. II (New York: 1981 and 1991), from *Interbike Directory*, various years, from United Nations, *Industrial Commodity Statistics Yearbook* (New York: various years), from *Bicycle Retailer and Industry News* world market summary data, various years, and from Otto Beaujon, managing editor, *Bike Europe* magazine, e-mail to author, 10 March 2006. Bicycle data are continually being revised as improved information becomes available. Thus the numbers here differ from those reported in earlier editions of *Vital Signs*.
2. *Bicycle Retailer and Industry News*, op. cit. note 1; Beaujon, op. cit. note 1.
3. *Bicycle Retailer and Industry News*, 1 March 2005.
4. John Crenshaw, "EU Passes Anti-Dumping Duties on China, Vietnam," *Bicycle Retailer and Industry News*, 15 August 2005.
5. Ibid.
6. John Crenshaw, "Europe Swims Against Tide of Asian Imports," *Bicycle Retailer and Industry News*, 1 September 2005.
7. "144 Percent: Mexico Doesn't Mess Around," *Bicycle Retailer and Industry News*, 1 November 2005.
8. Ibid.
9. Frank Jamerson, electric bicycle industry consultant, e-mail to author, 12 March 2006.
10. Ibid.
11. Worldwatch calculation based on estimate of 30 million bicycles sold in China from Crenshaw, op. cit. note 4.
12. Amy I. Zlot and Tom L. Schmid, "Relationships Among Community Characteristics and Walking and Bicycling for Transportation or Recreation," *American Journal of Health Promotion*, March/April 2005, pp. 314–17.
13. "Tapping the Market for Quality Bicycles in Africa,"

Sustainable Transport, e-update, December 2005.
14. Ibid.
15. "French City Launches Bike Loaner Program," *Bicycle Retailer and Industry News*, 1 October 2005.
16. Ibid.
17. Ibid.
18. Ibid.
19. Chris Lesser, "Green Bike Circulation is Booming at Arcata Library," *Bicycle Retailer and Industry News*, 1 April 2005.
20. Ibid.
21. Ibid.

AIR TRAVEL TAKES OFF AGAIN (pages 68–69)

1. International Civil Aviation Organization (ICAO), e-mail to author, 11 January 2005. All data from 2004 are provisional. ICAO states that provisional data are usually within plus or minus 2 percent of final numbers.
2. ICAO, op. cit. note 1; ICAO, "Scheduled Air Services Top 2 Billion Passengers in 2005," press release (Montreal: 15 December 2005).
3. ICAO, op. cit. note 1; ICAO, "World Air Passenger Traffic to Continue to Expand Through to 2007," press release (Montreal: 28 July 2005).
4. ICAO, op. cit. note 1; ICAO, op. cit. note 3.
5. ICAO, op. cit. note 3.
6. "Who Flies? Who Pays?" under Air Travel and Climate, Atmosfair, at www.atmosfair.de, viewed 25 January 2005.
7. Brandon Sewill, *Fly Now, Grieve Later: How to Reduce the Impact of Air Travel on Climate Change* (London: Aviation Environment Federation, 2005), p. 21. The calculation for Asia includes Japan.
8. ICAO, op. cit. note 3.
9. Sewill, op. cit. note 7, p. 7.
10. Ibid.
11. "Airbus Claims Victory in Jet War," *BBC News*, 17 January 2006.
12. Ibid.
13. "Boeing Plane Wins Distance Record," *BBC News*, 11 October 2005.
14. Engineering and Physical Sciences Research Council, "Climate Change and the Future of Air Travel," press release (London: 25 January 2005).
15. Ibid.
16. ICAO, "International Civil Aviation Day Calls for the Greening of Aviation," press release (Montreal: 30 November 2005).
17. Marla Dickerson, "Homegrown Fuel Supply Helps Brazil Breathe Easy," *Los Angeles Times*, 15 June 2005; "Brazil Launches Alcohol-powered Plane,"

Agence France-Presse, 16 March 2005.

18. Assad Kotaite, "Year-end Message from the President of the Council of the International Civil Aviation Organization" (Montreal: ICAO, December 2005).

19. ICAO, op. cit. note 16; ICAO, "Message from the Secretary General of the ICAO, Dr. Taïeb Cherif, for the Worldwide Celebration of International Civil Aviation Day on 7 December 2005" (Montreal: December 2005).

20. "U.K. Aviation Sector Promises Cleaner Planes," *Greenbiz.com*, 8 July 2005.

21. Ibid.

22. Sewill, op. cit. note 7, p. 34.

23. ICAO, "ICAO Council Adopts New Standards for Aircraft Emissions," press release (Montreal: 2 March 2005); Sewill, op. cit. note 7, p. 8.

24. Kotaite, op. cit. note 18.

25. "Climate Protection Projects," Atmosfair, at atmos fair.de, viewed 25 January 2006.

INTERNET AND CELL PHONE USE SOAR (pages 70–71)

1. International Telecommunications Union (ITU), "Mobile Cellular, Subscribers per 100 People," at www.itu.int/ITU-D/ict/statistics, viewed 22 March 2006. Five-year trend is a Worldwatch calculation.

2. Computer Industry Almanac, Inc., "China Tops Cellular Subscriber Top 15 Ranking," press release (Arlington Heights, IL: 26 September 2005).

3. Worldwatch calculation based on data from ITU, e-mail to author, 13 January 2006.

4. "Big Growth in Mobile Phone Sales," *BBC News*, 22 November 2005; Gartner, "Gartner Says Top Six Vendors Drive Worldwide Mobile Phone Sales to 21 Percent Growth in 2005," press release (Egham, U.K.: 28 February 2006).

5. Charlotte Windle, "China's Rich Fuel Mobile Revolution," *BBC News*, 5 December 2005; Computer Industry Almanac, Inc., op. cit. note 2.

6. Computer Industry Almanac, Inc., "Worldwide Internet Users Top 1 Billion in 2005," press release (Arlington Heights, IL: 26 September 2005).

7. ITU, "ICT Statistics," at www.itu.int/ITU-D/ict/sta tistics/ict, viewed 22 March 2006.

8. Worldwatch calculation based on data from ITU, op. cit. note 3.

9. Worldwatch calculation from Internet Systems Consortium, "ISC Domain Survey: Number of Internet Hosts," at www.isc.org, viewed 24 March 2006.

10. Christopher Rhoads, "In Threat to Internet's Clout, Some Are Starting Alternatives," *Wall Street Journal*, 19 January 2006; Hiawatha Bray, "China Creates Own Net Domains," *Boston Globe*, 1 March 2006.

11. Roland Buerk, "'Telephone Ladies' Connect Bangladesh," *BBC News*, 26 November 2005; Muhammad Yunus, *Grameen Bank at a Glance* (Dhaka, Bangladesh: Grameen Bank, February 2006).

12. Nicole Itano, "Africa's Cell Phone Boom Creates a Base for Low-cost Banking," *USA Today,* 28 August 2005.

13. Stefan Lovgren, "Can Cell-phone Recycling Help African Gorillas?" *National Geographic News*, 20 January 2006.

14. Christopher Cundy, "Concern Grows over Mobile Phone Disposal," *Plastics and Rubber Weekly*, 30 March 2006.

15. Lovgren, op. cit. note 13; recycling figure from Emily Lambert, "Use It Up, Wear It Out," *Forbes*, 14 March 2005.

16. Elizabeth Royte, "E-waste@large," *New York Times*, 27 January 2006; Lovgren, op. cit. note 13.

17. Lovgren, op. cit. note 13.

18. Basel Action Network, "Executive Summary: Are We Building High-tech Bridges or Waste Pipelines?" at ban.org/BANreports/10-24-05/index.htm, viewed 28 March 2006; hazardous waste examples from Silicon Valley Toxics Coalition, "High Tech Production," at www.svtc.org/hightech_prod/index.html.

19. Basel Action Network, "The Digital Dump: Exporting Re-use and Abuse to Africa," media release version (Seattle, WA: 24 October 2005), p. 4; Secretariat of the Basel Convention, "Introduction," at www.basel.int/pub/basics.html.

20. Silicon Valley Toxics Coalition and Basel Action Network, "Finally, A Responsible Way to Get Rid of That Old Computer!" press release (San Jose, CA, and Seattle, WA: 25 February 2003).

POPULATION CONTINUES TO GROW (pages 74–75)

1. U.S. Bureau of the Census, *International Data Base*, electronic database, Suitland, MD, updated August 2005.

2. Ibid.

3. Ibid.

4. Ibid.

5. U.N. Population Division, *World Population Prospects: The 2004 Revision* (New York: 2005).

6. Ibid.

7. Ibid.

8. Census Bureau, op. cit. note 1.

9. U.N. Population Division, op. cit. note 5.

10. Ibid.

11. U.N. Population Division, *World Urbanization Prospects: The 2003 Revision* (New York: 2004).

Notes

12. Ibid.
13. Tim Dyson and Pravin Visaria, "Migration and Urbanization: Retrospect and Prospects," in Tim Dyson, Robert Cassen, and Leela Visaria, eds., *21st Century India: Population, Economy, Human Development, and the Environment* (New York: Oxford University Press, 2004).
14. U.N. Population Division, op. cit. note 5.
15. Ibid.
16. Ibid.
17. Ibid.
18. Ibid.
19. Ibid.
20. Ibid.
21. Ibid.
22. John Leicester, "France Boosts Incentives for Having Kids," *The Guardian* (London), 21 September 2005.
23. U.N. Population Division, op. cit. note 5.
24. United Nations Population Fund, *State of World Population 2005* (New York: 2005).
25. Ibid.

HIV/AIDS THREATENS DEVELOPMENT (pages 76–77)

1. Estimates based on Joint United Nations Programme on HIV/AIDS (UNAIDS) and World Health Organization (WHO), *AIDS Epidemic Update* (Geneva: December 2005). UNAIDS estimates there are between 36.7 million and 45.3 million people currently living with HIV.
2. UNAIDS and WHO, op. cit. note 1.
3. UNAIDS and WHO, *AIDS Epidemic Update* (Geneva: various years). These estimates are based on data on the number of deaths published over the years by UNAIDS.
4. Ibid.; Figure 3 from Richard Cincotta, Robert Engelman, and Daniele Anastasion, *The Security Demographic: Population and Civil Conflict after the Cold War* (Washington, DC: Population Action International, 2003).
5. UNAIDS, *Caribbean: HIV/AIDS Statistics and Features, End of 2002 and 2004* (Geneva: December 2004).
6. UNAIDS and WHO, op. cit. note 1.
7. "WHO Says AIDS May Infect 10 mln in China by 2010," *Reuters*, 29 November 2005.
8. Ibid.
9. UNAIDS and WHO, op. cit. note 1, p. 33; "WHO Says AIDS May Infect 10 mln in China by 2010," op. cit. note 7.
10. Estimates based on UNAIDS and WHO, op. cit. note 3.
11. Fiona Fleck, "Developing Economies Shrink as AIDS Reduces Workforce," *BMJ*, 17 July 2004, p. 329.
12. International Labour Organization, *HIV/AIDS and Work: Global Estimates, Impacts and Response* (Geneva: 2004).
13. UNAIDS, "Fact Sheet on HIV/AIDS and Conflict" (Cophenhagen: August 2003).
14. UNICEF, "Childhood Under Threat: HIV/AIDS," in *The State of the World's Children 2005* (New York: 2005).
15. UNAIDS, *Population, Development and HIV/AIDS with Particular Emphasis on Poverty: The Concise Report* (New York: 2005).
16. WHO, "HIV Infection Rates Decreasing in Several Countries but Global Number of People Living with HIV Continues to Rise," press release (Geneva: 21 November 2005).
17. U.S. Centers for Disease Control and Prevention, "Data from the National HIV Prevention Conference Give a Clearer Picture of US HIV Status and Strategies to Confront It," *The 2005 National HIV Prevention Conference*, Atlanta, GA, 12–15 June 2005; UK Collaborative Group for HIV and STI Surveillance, "Mapping the Issues: HIV and Other Sexually Transmitted Infections in the United Kingdom: 2005," London, 24 November 2005.
18. Chinua Akukwe, "HIV/AIDS: Looming Funding Crisis," *Worldpress*, 27 July 2005.
19. Jennifer Kates, "An Overview: Funding for Global HIV/AIDS by Major Donor Governments," The Kaiser Family Foundation, July 2005, at www.kff.org/hivaids/7344./cfm, viewed 5 December 2005.
20. Akukwe, op. cit. note 18.

INFANT MORTALITY RATE FALLS AGAIN (pages 78–79)

1. United Nations Population Division, *World Population Prospects: The 2004 Revision* (New York: 2005).
2. Rates in 1950–91 from Hal Kane, "Infant Mortality Declining," in Worldwatch Institute, *Vital Signs 1992* (New York: W. W. Norton & Company, 1992), p. 78; rates in 1992–2005 from United Nations Population Division, op. cit. note 1.
3. World Health Organization (WHO), *World Health Report: Making Every Mother and Child Count* (Geneva: 2005), p. 9.
4. Ibid.
5. Ibid.
6. Centers for Disease Control and Prevention (CDC), "Infant, Neonatal, and Postneonatal Deaths, Percent of Total Deaths, and Mortality Rates for the 15 Leading Causes of Infant Death by Race and Sex," at www.cdc.gov/nchs/data/drs/LCWK7_2001.pdf, viewed 10 April 2006.

7. United Nations Population Division, op. cit. note 1.

8. Thierry Delvigne-Jean, "Integrated Approach Helps Cut Parent-to-Child HIV Transmission by Half in Cameroon," *UNICEF Newsline*, 6 June 2005.

9. United Nations Population Division, op. cit. note 1.

10. WHO, op. cit. note 3, p. 10.

11. Ibid., p. 104.

12. "Progress in Reducing Measles Mortality Worldwide 1999–2003," *Morbidity and Mortality Weekly Report*, 4 March 2005, pp. 200–03.

13. WHO, op. cit. note 3, p. xvii.

14. Ibid.

15. Ibid.

16. Ibid., p. 26.

17. CDC, "Eliminate Disparities in Infant Mortality," 5 April 2006, at www.cdc.gov/omh/AMH/factsheets/ infant.htm, viewed 10 April 2006; F. Edward Yazbak, "Concerns about the Infant Mortality Rate in the United States," *Red Flags,* 24 April 2005.

18. WHO, op. cit. note 3, p. 31.

19. UN Millennium Development Goals, at www.un.org/ millenniumgoals/index.html.

NUMBER OF VIOLENT CONFLICTS DROPS
(pages 82–83)

1. Arbeitsgemeinschaft Kriegsursachenforschung (AKUF), "Kaum Veränderungen im Kriegsgeschehen. Zahl der Kriegerischen Konflikte Leicht Zurückgegangen," press release (Hamburg, Germany: University of Hamburg, 16 December 2005); AKUF, "Das Kriegsgeschehen 2003 im Überblick," at www.sozialwiss.uni-hamburg.de/publish/Ipw/Akuf/ kriege_aktuell.htm, viewed 21 December 2004; Wolfgang Schreiber, "Daten und Tendenzen des Kriegsgeschehens 2004," in Wolfgang Schreiber, ed., *Das Kriegsgeschehen 2004. Daten und Tendenzen der Kriege und Bewaffneten Konflikte* (Wiesbaden, Germany: VS Verlag für Sozialwissenschaften, October 2005), pp. 12, 28. AKUF data are continually being revised as improved information becomes available. Thus the numbers here differ from those reported in earlier editions of *Vital Signs.*

2. AKUF, "Kaum Veränderungen im Kriegsgeschehen," op. cit. note 1; Schreiber, op. cit. note 1, p. 12.

3. AKUF, "Kaum Veränderungen im Kriegsgeschehen," op. cit. note 1; Schreiber, op. cit. note 1, p. 28.

4. AKUF, "Kaum Veränderungen im Kriegsgeschehen," op. cit. note 1.

5. Ibid.

6. "Memorandum of Understanding between the Government of Indonesia and the Free Aceh Movement," 15 August 2005, full text at Crisis Management Initiative, at www.cmi.fi/?content=aceh_project.

7. Michael Renner and Zoë Chafe, "Turning Disasters into Peacemaking Opportunities," in Worldwatch Institute, *State of the World 2006* (New York: W. W. Norton & Company, 2006), pp. 123–27.

8. Author's visit in Aceh, 15–23 December 2005.

9. Nils Petter Gleditsch et al., "Armed Conflict: 1946–2001: A New Dataset," *Journal of Peace Research*, vol. 39, no. 5 (2002), pp. 615–37; updated information in "The PRIO/Uppsala Armed Conflict Dataset. Armed Conflict, Version 3-2005b," released 6 September 2005, at www.prio.no/cwp/armedconflict. In Figure 2, "minor armed conflicts" involve at least 25 battle-related deaths in a year; if they involve more than 1,000 battle-related deaths during the course of the conflict, they are considered "intermediate conflicts." "Wars" are conflicts with at least 1,000 battle-related deaths in a given year. In addition to the conflicts shown in Figure 2, there are a number of "unclear cases" where uncertainty about the intensity of conflict defies easy categorization. The latest PRIO/Uppsala dataset does not include these unclear cases, and Figure 2 therefore understates the number of active conflicts.

10. B. Coghlan et al., "Mortality in the Democratic Republic of Congo: A Nationwide Survey," *The Lancet*, 7 January 2006, pp. 44–51.

11. Ibid.

12. Heidelberger Institut für Internationale Konfliktforschung (HIIK), *Conflict Barometer 2005* (Heidelberg, Germany: Institute for Political Science, University of Heidelberg, 2005), p. 1.

13. Ibid.

14. Ibid., p. 5.

15. Ibid.

MILITARY EXPENDITURES KEEP GROWING
(pages 84–85)

1. Unless otherwise noted, all monetary terms are expressed in constant 2005 dollars. Author's calculations based on Elisabeth Sköns et al., "Military Expenditure," in Stockholm International Peace Research Institute (SIPRI), *SIPRI Yearbook 2005: Armaments, Disarmament and International Security* (New York: Oxford University Press, 2005), p. 307; Rita Tulberg, "World Military Expenditure," *Bulletin of Peace Proposals*, no. 3–4, 1986; Bonn International Center for Conversion, *Conversion Survey* (Baden-Baden, Germany: Nomos Verlagsgesellschaft, annual), various editions; SIPRI, *SIPRI Yearbook* (New York: Oxford University Press, annual), various editions. Having been compiled from these different sources,

the time series presented in Figure 1 may entail some incompatibilities in terms of underlying methodologies. The intent is to provide a rough indication of longer-term trends.

2. Sköns et al., op. cit. note 1, p. 307.
3. Ibid., pp. 309–17.
4. Ibid., p. 309.
5. Ibid., pp. 310–14.
6. Ibid., p. 325.
7. Expressed in current dollar terms. Of the $347 billion in supplementary funds, $268 billion was allocated to the Department of Defense and $79 billion to the Departments of State, Homeland Security, and others; Sköns et al., op. cit. note 1, pp. 323–24.
8. David S. Cloud, "Billions Asked for Afghans and Iraqis," *New York Times*, 17 February 2006.
9. U.S. Department of Defense, Office of the Undersecretary of Defense (Comptroller), *National Defense Budget Estimates for FY 2006* (Washington, DC: April 2005), pp. 214–15.
10. Calculated by author, based on Sköns et al., op. cit. note 1, p. 318.
11. Ibid., p. 319; Elisabeth Sköns et al., "Military Expenditure," in SIPRI, *SIPRI Yearbook 2004: Armaments, Disarmament and International Security* (New York: Oxford University Press, 2004), pp. 312–13.
12. Elisabeth Sköns and Eamon Surry, "Arms Production," in SIPRI, *SIPRI Yearbook 2005*, op. cit. note 1, pp. 384–85. Original data in current dollar terms; translated into 2005 dollars by author.
13. Ibid., p. 384.
14. Sköns et al., op. cit. note 1, p. 307.
15. Ibid., p. 316.
16. Development assistance data from Organisation for Economic Co-operation and Development (OECD), "Aid Rising Sharply, According to Latest OECD Figures," undated, at www.oecd.org/dac, viewed 25 January 2006. Military spending–development assistance ratio based on calculations by author based on ibid. and on Sköns et al., op. cit. note 1, Tables 8A.1 and 8A.3.
17. OECD, op. cit. note 16.

PEACEKEEPING EXPENDITURES SET NEW RECORD (pages 86–87)

1. U.N. Department of Public Information (UNDPI), "United Nations Peacekeeping Operations. Background Note" (New York: 30 November 2005, and earlier editions); Worldwatch Institute database. All dollar amounts are in 2005 dollars.
2. U.N. Department of Peacekeeping Operations (UNDPKO), "Monthly Summary of Contributors," at www.un.org/Depts/dpko/dpko/contributors/index.htm, viewed 17 January 2006; personnel number also based on William Durch, Henry Stimson Center, Washington, DC, e-mail to author, 9 January 1996, and on Global Policy Forum, at www.globalpolicy.org/security/peacekpg/data/pkomctab.htm, viewed 22 December 2004.
3. UNDPI, op. cit. note 1; UNDPKO, op. cit. note 2.
4. UNDPI, "United Nations Political and Peace-Building Missions. Background Note" (New York: 31 October 2005).
5. Ibid.
6. Author's calculation, based on data from UNDPKO, op. cit. note 2. The percentage figures in this and the following paragraph refer to peacekeeping personnel excluding civilian staff.
7. Ibid.
8. Ibid.
9. Marc Lacey, "U.N. Forces Using Tougher Tactics to Secure Peace," *New York Times*, 23 May 2005.
10. UNDPKO, op. cit. note 2.
11. Ibid.
12. "Last U.N Troops Pack to Leave Sierra Leone as Mandate Expires," *New York Times*, 2 January 2006.
13. Calculated from UNDPI, op. cit. note 1.
14. Ibid.
15. Ibid.
16. Ibid.
17. United Nations, "UN Mission in Côte d'Ivoire Should Be Expanded and Extended for One Year, Annan Says," *UN News Service*, 13 January 2006.
18. U.N. Office for the Coordination of Humanitarian Affairs, "Côte d'Ivoire: Annan Wants More Peacekeepers on Ground," *IRINnews.org*, 5 January 2006.
19. Evelyn Leopold, "UN Envoy Cites Darfur Failure; Wants 20,000 Troops," *Reuters*, 14 January 2006.
20. UNDPI, op. cit. note 1.
21. December 2005 from Global Policy Forum, "US vs. Total Debt to the UN: 2005," at www.globalpolicy.org/finance/tables/core/un-us-05.htm, viewed 11 April 2006; December 2004 from "Status of Contributions to the Regular Budget, International Tribunals, Peacekeeping Operations and Capital Master Plan for the Biennium 2002–2003 as at 31 December 2004," from Mark Gilpin, Chief of United Nations Contributions Service, New York, letter to author, 17 January 2005.
22. Global Policy Forum, "Contributions Owing to the UN for Peacekeeping Operations: 2005," at www.globalpolicy.org/finance/tables/pko/due2005.htm, viewed 11 April 2006.
23. Global Policy Forum, op. cit. note 22.
24. Lacey, op. cit. note 9.

25. International Commission on Intervention and State Sovereignty, *The Responsibility to Protect* (Ottawa, ON, Canada: 2001).

26. United Nations, Department of Peacekeeping Operations, "Fatalities by Year up to 29 Dec 2005," at www.un.org/Depts/dpko/fatalities/totals_annual.htm.

27. Renata Dwan and Sharon Wiharta, "Multilateral Peace Missions: Challenges of Peace-Building," in Stockholm International Peace Research Institute, *SIPRI Yearbook 2005* (New York: Oxford University Press, 2005), pp. 139–98; International Institute for Strategic Studies (IISS), "The 2005 Chart of Armed Conflict," wall chart distributed with IISS, *The Military Balance 2005–2006* (London: Oxford University Press, 2005); Worldwatch Institute database.

28. Worldwatch estimates, based primarily on data from Dwan and Wiharta, op. cit. note 27, and on IISS, "The 2005 Chart," op. cit. note 27.

29. Dwan and Wiharta, op. cit. note 27.

30. Aceh Monitoring Mission (AMM), at www.aceh-mm .org; AMM representatives, Sigli, Indonesia, discussions with author, 18 December 2005.

GLOBAL ECOSYSTEMS UNDER MORE STRESS (pages 92–93)

1. Millennium Ecosystem Assessment (MA), *Ecosystems and Human Well-being: Synthesis* (Washington, DC: Island Press, 2005), p. viii.

2. Ibid., p. 1.

3. Ibid.

4. Ibid., pp. 2–5.

5. Ibid., p. 1.

6. Ibid.

7. MA, *Living Beyond Our Means: Natural Assets and Human Well-Being: Statement from the Board* (Washington, DC: World Resources Institute, 2005), p. 5.

8. World Wide Fund for Nature (WWF), U.N. Environment Programme's World Conservation Monitoring Centre (WCMC), and Global Footprint Network, *Living Planet Report 2004* (Gland, Switzerland: WWF, 2004).

9. Robert Prescott-Allen, *The Wellbeing of Nations* (Washington, DC: Island Press, 2001), pp. 303–06.

10. Ibid., p. 59.

11. WWF, WCMC, and Global Footprint Network, op. cit. note 8. The Ecological Footprint is a conservative estimate of human usage of resources as it only includes renewable resources and does not consider the effects of acid rain or the usage of heavy metals, radioactive materials, or persistent synthetic chemicals on Earth's biocapacity. The measure also does not set aside any biocapacity for wild species.

12. WWF, WCMC, and Global Footprint Network, op. cit. note 8.

13. Ibid., with 2002 update from Global Footprint Network, *National Footprint and Biocapacity Accounts, 2005 Edition* (Copenhagen: European Environment Agency, 2005). The productivity of "global hectares" is an average based on the productivity of ecosystems used by humans.

14. Global Footprint Network, op. cit. note 11.

15. WWF, WCMC, and Global Footprint Network, op. cit. note 8, and Global Footprint Network, op. cit. note 11.

16. Global Footprint Network, op. cit. note 11.

17. Calculations based on ibid.; U.S. Bureau of the Census, *International Data Base*, electronic database, Suitland, MD, updated August 2005.

18. Calculations based on Global Footprint Network, op. cit. note 11.

19. Population from Census Bureau, op. cit. note 15; economic growth from Christopher Flavin and Gary Gardner, "China, India, and the New World Order," in Worldwatch Institute, *State of the World 2006* (New York: W. W. Norton & Company, 2006), pp. 3–23; consumption from Gary Gardner, Erik Assadourian, and Radhika Sarin, "The State of Consumption Today," in Worldwatch Institute, *State of the World 2004* (New York: W. W. Norton & Company, 2004), pp. 3–21.

20. MA, op. cit. note 1, pp. 18–24.

21. World Economic Forum, *Global Governance Initiative Annual Report 2006* (Washington, DC: 2006).

22. Felicity Barringer, "United States Ranks 28th on Environment, a New Study Says," *New York Times*, 23 January 2006.

23. Yale Center for Environmental Law & Policy and Center for International Earth Science Information Network, *Pilot 2006 Environmental Performance Index Report* (New Haven, CT: Yale Center for Environmental Law & Policy, 2006), p. 10.

24. Ibid., p. 3.

25. Ibid., pp. 3, 21.

26. Ibid., p. 18.

27. Ibid., p. 20.

CORAL REEF LOSSES INCREASING (pages 94–95)

1. Clive Wilkinson, ed., *Status of Coral Reefs of the World: 2004. Volume 1* (Townsville, Queensland, Australia: Australian Institute of Marine Science/Global Coral Reef Monitoring Network, 2004), p. 9.

2. Ibid.

3. Reef area figures vary from data in *Vital Signs 2001* due to new reef discoveries and revised estimates.

Table 1 based on the following: area covered from United Nations Environment Programme/World Conservation Monitoring Centre (UNEP/WCMC), *World Atlas of Coral Reefs* (Cambridge, U.K.: 2001); share destroyed and trends from Wilkinson, op. cit. note 1; "Fiji Chiefs Create Marine Sanctuaries on World's Third Largest Reef," *Environment News Service*, 4 November 2005.

4. Half-billion from Dirk Bryant et al., *Reefs at Risk* (Washington, DC: World Resources Institute (WRI), 1998), p. 10; 100 countries from Reef Relief, "Map of the World's Coral Reefs," at www.reefrelief.org/coral reef/maps/map.html, viewed 19 December 2005.

5. IUCN–World Conservation Union, "Healthy Corals Fared Best Against Tsunami," press release (Gland, Switzerland: 15 December 2005).

6. Wilkinson, op. cit. note 1, p. iii.

7. Mark Shwartz, "Scientists Deliver Plan For Rescuing America's Coral Reefs," *Stanford Report*, 28 March 2005.

8. UNEP/WCMC, International Coral Reef Action Network (ICRAN), and IUCN, *In the Front Line: Shoreline Protection and Other Ecosystem Services from Mangroves and Coral Reefs* (Nairobi: 2006).

9. Area covered from "Caribbean Reefs Bleached by Warm Water," *Associated Press*, 3 November 2005; complex and productive from Jamie K. Reaser, Rafe Pomerance, and Peter O. Thomas, "Coral Bleaching and Global Climate Change: Scientific Findings and Policy Recommendations," *Conservation Biology*, 5 October 2000, p. 1501.

10. Reaser, Pomerance, and Thomas, op. cit. note 9.

11. "Scientists to Explore Newly Discovered Oman Reef," *Reuters*, 27 October 1999; HIV from Paul Majendie, "Great Barrier Reef Needs Saving From Fame," *Reuters*, 14 September 2000.

12. Wilkinson, op. cit. note 1; Magnus Nyström, Carl Folke, and Fredrik Moberg, "Coral Reef Disturbance and Resilience in a Human-Dominated Environment," *Trends in Ecology and Evolution*, 10 October 2000, p. 413; "Longest Reef Escapes Oil Spill, But Dugongs Dying," *Environment News Service*, 18 May 1999; "Grounded Ship Damages Part of Australian Coral Reef," *Reuters*, 6 November 2000.

13. Wilkinson, op. cit. note 1, p. 28.

14. Andrew W. Bruckner, "New Threat to Coral Reefs: Trade in Coral Organisms," *Issues in Science and Technology*, fall 2000.

15. Gregor Hodgson and Jennifer Liebeler, *The Global Coral Reef Crisis: Trends and Solutions 1997–2001* (Los Angeles: Reef Check, August 2002).

16. Wilkinson, op. cit. note 1, p. 27.

17. Lauretta Burke, Liz Selig, and Mark Spalding, *Reefs at Risk in Southeast Asia* (Washington, DC: WRI, 2002), p. 28.

18. John Ryan, "Blast Fishing Competes With Reef Conference," *Environmental News Network*, 28 October 2000.

19. Bryant et al., op. cit. note 4, p. 15.

20. Andrew Darby, "Great Barrier Reef Crisis Growing," *Environment News Service*, 14 January 1999.

21. Wilkinson, op. cit. note 1; exacerbate from R.W. Buddemeier, J.A. Kleypas, and R. Aronson, *Coral Reefs and Global Climate Change. Potential Contributions of Climate Change to Stresses on Coral Reef Ecosystems*, prepared for the Pew Center for Global Climate Change (Arlington, VA: 2004).

22. Early indicator from Luitzen Bijlsma, "Coastal Zones and Small Islands," in Robert T. Watson et al., eds., *Climate Change 1995. Impacts, Adaptations and Mitigation of Climate Change: Scientific-Technical Analyses* (New York: Cambridge University Press, 1996), pp. 303–04.

23. Ove Hoegh-Guldberg, *Climate Change: Coral Bleaching and the Future of the World's Coral Reefs* (Amsterdam: Greenpeace International, 1999), p. 1.

24. Ibid, p. 5.

25. Reaser, Pomerance, and Thomas, op. cit. note 9, p. 1503.

26. Worst on record and 16 percent from Wilkinson, op. cit. note 1, p. 1; thousand years and 40 meters from University of California–Los Angeles, "New Global Research Shows Coral Reefs Remain in Jeopardy; UCLA Scientist Announces Southern California Marine Ecology Health Study," press release (Los Angeles: 20 April 2000).

27. Estimate of 90 percent from IUCN, *Coral Reef Resilience and Resistance to Bleaching* (Gland, Switzerland: 2005); damages from Reaser, Pomerance, and Thomas, op. cit. note 9, p. 1508.

28. Wilkinson, op. cit. note 1, p. 8.

29. Ibid, p. 7.

30. IUCN, op. cit. note 27.

31. "Caribbean Reefs Bleached by Warm Water," *Associated Press*, 3 November 2005.

32. Wilkinson, op. cit. note 1, pp. 24–25.

33. IUCN, op. cit. note 27.

34. Wilkinson, op. cit. note 1, pp. 12–13.

35. United States Coral Reef Task Force, *The National Action Plan to Conserve Coral Reefs* (Washington, DC: 2000), p. 20.

BIRDS REMAIN THREATENED (pages 96–97)

1. BirdLife International, *State of the World's Birds 2004: Indicators for Our Changing World* (Cambridge, U.K.:

2004), p. 8, with updates from Stuart Butchart, discussion with author, 15 February 2006.

2. Çağan H. Sekercioğlu, Gretchen C. Daily, and Paul R. Ehrlich, "Ecosystem Consequences of Bird Declines," *Proceedings of the National Academy of Sciences*, 28 December 2004, pp. 18042–47.

3. José Maria Cardoso da Silva and Marcelo Tabarelli, "Tree Species Impoverishment and the Future Flora of the Atlantic Forest of Northeast Brazil" (letter), *Nature*, 2 March 2000, pp. 72–74.

4. Milton Friend, Robert G. McClean, and Joshua F. Dein, "Disease Emergence in Birds: Challenges for the Twenty-First Century," *The Auk*, April 2001, pp. 290–303; Sekercioglu, Daily, and Ehrlich, op. cit. note 2.

5. BirdLife International, op. cit. note 1, p. 8.

6. Millennium Ecosystem Assessment, *Ecosystems and Human Well-being: Synthesis* (Washington, DC: Island Press, 2005).

7. Birdlife International, op. cit. note 1, p. 32.

8. Noel Snyder et al., eds., *Parrots, Status Survey and Conservation Action Plan 2000–2004* (Gland, Switzerland, and Cambridge, U.K.: IUCN–World Conservation Union, 2000), p. x, with updates from Jamie Gilardi, director, World Parrot Trust, discussion with author, 2 February 2006.

9. Peter Germann and Peter Holland, "Fragmented Ecosystems: People and Forests in the Mountains of Switzerland and New Zealand," *Mountain Research and Development*, November 2001, pp. 382–91.

10. Snyder et al., op. cit. note 8.

11. BirdLife International, op. cit. note 1.

12. Snyder et al., op. cit. note 8.

13. Ibid.

14. BirdLife International, op. cit. note 1, p. 42.

15. Royal Society for the Protection of Birds, at www.rspb.org.uk/international/albatross_appeal/index.asp, viewed 15 March 2006.

16. BirdLife International, op. cit. note 1, p. 44.

17. Ibid.

18. Pacific Regional Environment Programme, at www.sprep.org.ws/topic/Invasiv.htm, viewed 15 March 2006.

19. BirdLife International, op. cit. note 1, p. 44.

20. Ibid, p. 40.

21. Ibid.

22. Ibid.

23. Mike Hoffman, Conservation International, discussion with author, 13 February 2006; Royal Society for the Protection of Birds, at www.rspb.org.uk/countryside/farming/advice/birdsonfarms/song thrush/need.asp, viewed 15 March 2006.

24. BirdLife International, op. cit. note 1, p. 40; Björn Helander et al., "The Role of DDE, PCB, Coplanar PCB and Eggshell Parameters for Reproduction in the White-tailed Sea Eagle (*Haliaeetus albicilla*) in Sweden," *Ambio*, vol. 31, no. 5 (2002), pp. 386–403.

25. Peter A. Cotton, "Avian Migration Phenology and Global Climate Change," *Proceedings of the National Academy of Sciences*, 14 October 2003, pp. 12219–22.

26. Ibid.

27. Jeff Price and Patricia Glick, *The Birdwatchers' Guide to Global Warming* (Washington, DC: American Bird Conservancy and National Wildlife Federation, 2002).

28. Randi Doeker, The Birds & Buildings Forum, discussion with author, 2 February 2006.

29. BirdLife International, op. cit. note 1, p. 38.

30. Rachel Bristol, science coordinator, Nature Seychelles, discussion with author, 2 February 2006.

31. J. E. M. Baillie, C. Hilton-Taylor, and S. N. Stuart, eds., *2004 IUCN Red List of Threatened Species* (Gland, Switzerland, and Cambridge, U.K.: IUCN–World Conservation Union, 2004).

PLANT DIVERSITY ENDANGERED (pages 98–99)

1. IUCN Species Survival Commission, *2004 IUCN Red List of Threatened Species*, at www.iucnredlist.org.

2. E. N. Lughadha et al., "Measuring the Fate of Plant Diversity: Towards a Foundation for the Future Monitoring and Opportunities for Urgent Action," *Philosophical Transactions: Biological Sciences (Royal Society of London)*, 28 February 2005, pp. 359–72.

3. IUCN Species Survival Commission, op. cit. note 1.

4. Ibid.

5. T. H. Ricketts et al., "Pinpointing and Preventing Imminent Extinctions," *Proceedings of the National Academy of Sciences*, 20 December 2005, pp. 18497–501.

6. U.N. Food and Agriculture Organization, *The State of the World's Plant Genetic Resources for Food and Agriculture* (Rome: 1997); Convention on Biological Diversity, *Second Global Biodiversity Outlook: Draft Executive Summary* (Montreal, Canada: 2005).

7. P. M. Vitousek et al., "Human Appropriation of the Products of Photosynthesis," *BioScience*, June 1986, pp. 368–73; S. Rojstaczer, S. M. Sterling, and N. J. Moore, "Human Appropriation of Photosynthesis Products," *Science*, 21 December 2001, pp. 2549–52; M. L. Imhoff et al., "Global Patterns in Human Consumption of Net Primary Productivity," *Nature*, 24 June 2004, pp. 870–73.

8. R. V. O'Neill and J. R. Kahn, "*Homo economus* as a Keystone Species," *BioScience*, April 2000, pp. 333–37.

9. Millennium Ecosystem Assessment (MA), *Ecosystems and Human Well-being: Biodiversity Synthe-

Notes

sis (Washington, DC: World Resources Institute, 2005), pp. 22–25, 28–29.

10. J. van Ruijven and F. Berendse, "Diversity-productivity Relationships: Initial Effects, Long-term Patterns, and Underlying Mechanisms," *Proceedings of the National Academy of Sciences*, 18 January 2005, pp. 695–700; B. Schmid, J. Joshi, and F. Schläpfer, "Empirical Evidence for Biodiversity-Ecosystem Functioning Relationships," in A. Kinzig, D. Tilman, and S. W. Pacala, eds., *The Functional Consequences of Biodiversity: Empirical Progress and Theoretical Extensions* (Princeton, NJ: Princeton University Press, 2002), pp. 120–50; D. Tilman et al., "Plant Diversity and Composition: Effects on Productivity and Nutrient Dynamics of Experimental Grasslands," in M. Loreau, S. Naeem, and P. Inchausti, eds., *Biodiversity and Ecosystem Functioning: Synthesis and Perspectives* (Oxford: Oxford University Press, 2002), pp. 21–35; D. Tilman et al., "Experimental and Observational Studies of Diversity, Productivity, and Stability," in Kinzig, Tilman, and Pacala, op. cit. this note, pp. 42–70; Lughadha et al., op. cit. note 2.

11. MA, op. cit. note 9, pp. 8–10.

12. Ibid.

13. Ibid., p. 60–61.

14. Ibid., p. 61.

15. W. Thuiller et al., "Climate Change Threats to Plant Diversity in Europe," *Proceedings of the National Academy of Sciences*, 7 June 2005, pp. 8245–50.

16. Convention on Biological Diversity, at www.biodiv.org.

17. MA, op. cit. note 9, p. 77.

18. Ibid.

19. UN Millennium Development Goals, at www.un.org/millenniumgoals/index.html.

20. MA, op. cit. note 9, p. 81.

DISAPPEARING MANGROVES LEAVE COASTS AT RISK (pages 100–01)

1. U.N. Food and Agriculture Organization (FAO), "Twenty Percent of the World's Mangroves Lost over the Last 25 Years: Rate of Deforestation Slowing, But Still Cause for Alarm," press release (Rome: 9 November 2005); coast vulnerability from F. Danielsen et al., "Coastal Vegetation and the Asian Tsunami: Response," *Science*, 6 January 2006, pp. 37–38.

2. Mette Løyche Wilkie, FAO, e-mail to author, 31 January 2006.

3. FAO, op. cit.note 1.

4. International Tropical Timber Organization (ITTO), *Mangrove Work Plan 2002–2006* (June 2002), pp. 1–2.

5. FAO, *Mangrove Forest Management Guidelines*, FAO Forestry Paper No. 117 (Rome: 1994).

6. ITTO, op. cit. note 4, pp. 1–2.

7. International Society for Mangrove Ecosystems (ISME), *ISME Mangrove Plan for the Sustainable Management of Mangroves 2004–2009* (Okinawa, Japan: 2004), p. 5.

8. FAO, op. cit. note 1; country figures from Earth Trends, World Resources Institute, at earthtrends.wri.org.

9. FAO, op. cit. note 1.

10. ITTO, op. cit. note 4, p. 2; crabs and oysters from World Wide Fund for Nature (WWF) International, "Deforestation Threatens the Cradle of Reef Diversity," Feature Story (Gland, Switzerland: 2 December 2004); birds from Alfredo Quarto, "The Mangrove Forest," background paper (Gland, Switzerland: Ramsar Convention on Wetlands Secretariat, 1997).

11. P. Mumby et al., "Mangroves Enhance the Biomass of Coral Reef Fish Communities in the Caribbean," *Nature*, 5 February 2004, pp. 533–36.

12. ITTO, op. cit. note 4, p. 2; siltation from ISME, op. cit. note 7, p. 5.

13. ISME, op. cit. note 7, p. 5.

14. ITTO, op. cit. note 4, p. 2.

15. WWF International, op. cit. note 10.

16. Mangrove Action Project, "How Much is a Mangrove Forest Worth?" brochure (Port Angeles, WA: undated).

17. ITTO, op. cit. note 4, p. 2; WWF International, op. cit. note 10.

18. WWF International, op. cit. note 10.

19. Mangrove Action Project, op. cit. note 16.

20. ITTO, op. cit. note 4, p. 2.

21. Mangrove Action Project, op. cit. note 16.

22. FAO, "Tsunami Reconstruction: Tsunami Mitigation by Mangroves and Coastal Forests," at www.fao.org/forestry/site/27285/en, viewed 31 January 2006.

23. Erika Check, "Natural Disasters: Roots of Recovery," *Nature*, 15 December 2005, pp. 910–11.

24. F. Dahdouh-Guebas, "Mangrove Forests and Tsunami Protection," *McGraw-Hill Yearbook of Science and Technology* (New York: McGraw-Hill Professional, 2006), p. 190.

25. F. Danielsen et al., "The Asian Tsunami: A Protective Role for Coastal Vegetation," *Science*, 28 October 2005, p. 643; FAO, "Tsunami Reconstruction: Past and Current Extent of Mangroves in the Affected Countries," at www.fao.org/forestry/site/27285/en, viewed 31 January 2006.

26. J. Primavera, "Mangroves, Fishponds, and the Quest for Sustainability," *Science*, 7 October 2005, p. 59.

27. Ibid.

28. WWF International, "Eastern Africa Marine Ecoregion: Causes and Effects of Coastal Degradation,"

at www.panda.org/about_wwf/where_we_work/
africa/what_we_do/eastern_africa/our_solutions/
eame/area/index/causes_effects/index.cfm, November 2005.

29. "Indonesia Sees Years of Work to Repair Ecosystem Damage," *International Herald Tribune*, 1 February 2005.

30. High Court of Judicature at Bombay, Writ Petition (Lodging) No. 3246 of 2004, 6 October 2005.

31. Isabelle Louis, "After the Tsunami: Helping People by Helping the Environment," *International Herald Tribune*, 20 January 2005.

32. IUCN–World Conservation Union, "Mangroves for the Future: Reducing Vulnerability and Improving Livelihoods After the Tsunami," brochure (Gland, Switzerland: undated).

33. Danielsen et al., op. cit. note 25.

34. WWF International, *Saving Pakistan's Green Gold* (Gland, Switzerland: DGIS-ICD Programme, 2004).

35. Check, op. cit. note 23.

36. Ibid.

37. Mangrove Action Project, "WWF Wrongly Endorses Shrimp Farming in Aceh to Boost Tsunami Recovery!" e-mail to author, 9 March 2006.

38. Primavera, op. cit. note 26, p. 59.

39. WWF International, op. cit. note 10.

40. Mangrove Action Project, "Wal-Mart and Darden Restaurants Announce Future Sourcing of 'Certified' Farm-raised Shrimp: Will Consumers Be Served 'Green' Shrimp, or a Green-wash?" press release (Port Angeles, WA: 29 January 2006).

DEFORESTATION CONTINUES (pages 102–03)

1. Worldwatch calculation based on data in "Change in Extent of Forest and Other Wooded Land 1990–2005," in U.N. Food and Agriculture Organization (FAO), *Global Forest Resources Assessment* (Rome: 2005).

2. Ibid.

3. Ibid.

4. Ibid.

5. Ibid.

6. Ibid.

7. Ibid.

8. Ibid.

9. Tamara Stark and Sze Pang Cheung, *Sharing the Blame: Global Consumption and China's Role in Ancient Forest Destruction* (Amsterdam: Greenpeace International and Greenpeace China, 2006).

10. Dirk Bryant et al., *The Last Frontier Forests: Ecosystems and Economies on the Edge* (Washington, DC: World Resources Institute, 1997), p. 43.

11. FAO, *Global Forest Resources Assessment: Key Findings* (Rome: 2005).

12. Ibid.

13. Ibid.

14. Helmut J. Geist and Eric F. Lambin, *What Drives Tropical Deforestation?* LUCC Report Series No. 4 (Louvain-la-Neuve, Belgium: Land Use and Land Cover Change Project, 2001), p. 1.

15. David Kaimowitz et al., *Hamburger Connection Fuels Amazon Destruction* (Jakarta: Center for International Forestry Research, 2004), p. 2.

16. Ibid.

17. Ibid, pp. 3–4.

18. Millennium Ecosystem Assessment, *Ecosystems and Human Well-being: Current State and Trends, Volume 1* (Washington, DC: Island Press, 2005), pp. 600–07.

19. FAO, op. cit. note 11.

20. FAO, "Incentives to Curb Deforestation Needed to Counter Climate Change," press release (Rome: 9 December 2005).

21. C. M. Gough et al., "The Legacy of Forest Harvest and Burning on Ecosystem Carbon Storage in the Northern Midwest, USA," presented to the fall meeting of the American Geophysical Union, 5–9 December 2005.

22. Rhett A. Butler, "Rainforests Worth $1.1 Trillion for Carbon Alone in 'Coalition' Nations," at www.mongabay.com, 27 November 2005.

23. Ibid.

24. Mickey Lam, "Rainforest Nations Seek Credit for Preservation," *Epoch Times*, 12 December 2005.

25. Ibid.

26. Worldwatch calculation based on data in "Information on Certified Forest Sites Endorsed by Forest Stewardship Council (FSC)," at www.certified-forests.org.

27. Ibid.

28. Worldwatch calculation based on data in "Information on Certified Forest Sites," op. cit. note 26, and in FAO, op. cit. note 1.

29. Ibid.

30. Ibid.

GROUNDWATER OVERDRAFT PROBLEM PERSISTS (pages 104–05)

1. S. S. D. Foster and P. J. Chilton, "Groundwater: The Processes and Global Significance of Aquifer Degradation," *Philosophical Transactions: Biological Sciences*, 29 December 2003, pp. 1957–72.

2. Leonard F. Konikow and Eloise Kendy, "Groundwater Depletion: A Global Problem," *Hydrogeology Journal*, March 2005, pp. 317–20; Table 1 from Mark

Notes

W. Rosegrant, Ximing Cai, and Sarah A. Cline, *World Water and Food to 2025* (Washington, DC: International Food Policy Research Institute and International Water Management Institute, 2002).

3. J. R. Bartolino and W. L. Cunningham, "Ground-Water Depletion Across the Nation," fact sheet (Reston, VA: U.S Geological Survey (USGS), November 2003).

4. Maude Barlow, *Blue Gold: The Global Water Crisis and the Commodification of the World's Water Supply* (San Francisco: International Forum on Globalization, 1999).

5. Joanna Kidd, Lura Consulting, "Groundwater: A North American Resource," Discussion Paper for Expert Workshop on Freshwater in North America, 21 January 2002.

6. Ibid.

7. Barlow, op. cit. note 4.

8. Ibid.

9. Ibid.

10. Central Ground Water Board, Ministry of Water Resources, Government of India, at cgwb.gov.in/dist _names_morethen20_pre.htm.

11. Foster and Chilton, op. cit. note 1.

12. China Geological Survey, *China Groundwater Resources and Environment Survey Report* (Beijing: March 2005).

13. Ibid.

14. D. L. Galloway, D. R. Jones, and S. E. Ingebritsen, eds., "Land Subsidence in the United States," Circular 1182 (Reston, VA: USGS, 1999).

15. National Research Council, *Investigating Groundwater Systems on Regional and National Scales* (Washington, DC: National Academy Press, 2000).

16. Igor S. Zektser, *Groundwater and the Environment: Applications for the Global Community* (Boca Raton, FL: CRC Press, 2000).

17. Luis E. Marin, "Perspectives on Mexican Ground Water Resources, 2002," at geoinf.igeolcu.unam.mx/ rda/memb/lmarin/perspectives.pdf.

18. Ibid.

19. China Geological Survey, op. cit. note 12.

20. Ibid.

21. National Research Council, op. cit. note 15.

22. P. M. Barlow, "Ground Water in Freshwater Saltwater Environments of the Atlantic Coast," Circular 1262 (Reston, VA: USGS, 2003); C. J. Taylor and W. M. Alley, "Ground-water Level Monitoring and the Importance of Long-term Water-level Data," Circular 1217 (Reston, VA: USGS, 2001).

23. Central Ground Water Board, op. cit. note 10.

24. China Geological Survey, op. cit. note 12.

25. Ibid.

26. S. Zektser, H. A. Loáiciga, and J. T. Wolf, "Environmental Impacts of Groundwater Overdraft: Selected Case Studies in the Southwestern United States," *Environmental Geology*, February 2005, pp. 396–404.

27. "Water Diversion Project Ready for Construction in 2002," *People's Daily*, 15 November 2001.

28. Centre for Science and Environment, "A Look at India's Water Harvesting Practices," at rainwaterhar vesting.org/Rural/Rural.htm.

REDUCING MERCURY POLLUTION (pages 106–07)

1. Health effects from exposure during development from National Academy of Sciences/National Research Council, Committee on the Toxicological Effects of Methylmercury, *Toxicological Effects of Methylmercury* (Washington, DC: National Academy Press, 2000).

2. Ibid.; Jyrki K. Virtanen et al., "Mercury, Fish Oils, and Risk of Acute Coronary Events and Cardiovascular Disease, Coronary Heart Disease, and All-Cause Mortality in Men in Eastern Finland," *Arteriosclerosis, Thrombosis, and Vascular Biology*, January 2005, pp. 228–33.

3. Contamination of the Arctic documented in Marla Cone, *Silent Snow: The Slow Poisoning of the Arctic* (New York: Glover Press, 2005).

4. Mercury concentrations in animals from Henrik Skov et al., *Fate of Mercury in the Arctic* (Denmark: National Environmental Research Institute, 2004); increases discussed in R. Wagemann et al., "Overview and Regional and Temporal Differences of Heavy Metals in Arctic Whales and Ringed Seals in the Canadian Arctic," *Science of the Total Environment*, vol. 186 (1996), pp. 41–67, and in D. Muir et al., "Temporal Trends of Persistent Organic Pollutants and Metals in Ringed Seals from the Canadian Arctic," in *Synopsis of Research Conducted under the 2000/01 Northern Contaminants Program* (Ottawa, ON, Canada: Indian and Northern Affairs Canada, 2001).

5. Trends in mercury demand reported by P. Maxson, *Mercury Flows in Europe and the World: The Impact of Decommissioned Chlor-Alkali Plants*, report for the European Commission–DG Environment (Brussels: 2004).

6. Ibid.

7. Ibid.

8. Ibid.

9. Ibid.

10. Ibid.

11. Ibid.

12. Shen Yingwa, deputy director, Chemical Registra-

tion Center, State Environmental Protection Agency, China, discussions with Linda Greer, 14 October 2005.

13. Authors' estimates, based on currently available alternative technologies.

14. Council of the European Union, "Council Conclusions on the Community Strategy Concerning Mercury," 2670th Environment Council meeting (Luxembourg: 24 June 2005).

15. Shen, op. cit. note 12.

REGIONAL DISPARITIES IN QUALITY OF LIFE PERSIST (pages 110–11)

1. United Nations Development Programme (UNDP), *Human Development Report 2005* (New York: Oxford University Press, 2005), p. 20.

2. Ibid.

3. Ibid.

4. U.N. Food and Agriculture Organization (FAO), "Number of Undernourished People," *FAOSTAT Statistical Database*, at apps.fao.org, viewed 4 April 2006.

5. "Water, Sanitation, and Nutritional Status," U.N. Millennium Indicators Database, at millenniumindicators.un.org, viewed April 2006.

6. UNDP, op. cit. note 1, p. 19.

7. Ibid.

8. Global Alliance for Vaccines, "Progress and Achievements," at www.vaccinealliance.org/resources/ FS_Progress___Achievements_en_Jan05.pdf

9. UNDP, op. cit. note 1, p. 20.

10. Ibid.

11. U.N. Statistics Division, "Statistics and Indicators on Men and Women," at unstats.un.org, viewed April 2006.

12. Center for International Development and Conflict Management, cited in UNDP, op. cit. note 1, p. 20.

13. UNDP, op. cit. note 1, p. 20.

14. Ibid.

15. United Nations, *The Millennium Development Goals Report* (New York: 2005), p. 7.

16. UNDP, op. cit. note 1, p. 23.

17. United Nations, op. cit. note 15, p. 6.

18. Poverty rate from UNDP, op. cit. note 1, p. 21; number of hungry from FAO, op. cit. note 4.

19. World Health Organization (WHO), *WHO Report on Infectious Diseases: Removing Obstacles to Healthy Development*,(Geneva: 1999).

20. UNDP, op. cit. note 1, p. 55.

21. WHO, *Evolution of Global Suicide Rates, 1950–2000* (Geneva: 2002).

22. U.S. Census Bureau, "Poverty: Trends for Selected Groups," at www.census.gov, revised 30 August 2005; UNDP, op. cit. note 1, p. 227.

23. Ronald Inglehart and Hans-Dieter Klingemann, "Genes, Culture, Democracy, and Happiness," in E. Diener and E. M. Suh, eds., *Culture and Subjective Well-Being* (Cambridge, MA: The MIT Press, 2000), p. 171.

24. U.S. National Institutes of Health, "Obesity Threatens to Cut U.S. Life Expectancy," press release (Bethesda, MD: 16 March 2005).

25. Rut Veenhoven, *Trend Average Happiness in Nations 1946–2004: How Much People Like the Life They Live—World Database of Happiness* (Rotterdam, Netherlands: Erasmus University, 2005).

26. Ibid.

27. FAO, "Impact of Climate Change, Pests and Diseases on Food Security and Poverty Reduction," *Special Event Background Document for the 31st Session of the Committee on World Food Security*, Rome, 23–26 May 2005.

28. Millennium Ecosystem Assessment, *Ecosystems and Human Well-being: Synthesis* (Washington, DC: Island Press, 2005).

LANGUAGE DIVERSITY DECLINING (pages 112–13)

1. "Babel Runs Backwards—Endangered Languages," *The Economist* (U.S. edition), 1 January 2005.

2. The exact number of spoken languages in the world is a hotly debated topic among linguists. Most experts agree that there are nearly 7,000 languages in use today. Most reference books since the 1980s estimate that 6,000–7,000 languages currently exist, but in the last 10 years estimates have varied between 3,000 and 10,000; David Crystal, *Language Death* (Cambridge, UK: Cambridge University Press, 2000), p. 3. The 15th edition of Ethnologue lists 6,912 languages and the *International Encyclopedia of Linguistics* lists 6,300 living languages: Raymond G. Gordon, Jr., ed., *Ethnologue: Languages of the World*, 15th ed. (Dallas, TX: Summer Institute of Linguistics International, 2005); William Bright, *International Encyclopedia of Linguistics* (New York: Oxford University Press, 1992).

3. Gordon, op. cit. note 2; UNESCO Web site, at portal.unesco.org/culture, updated November 2005.

4. Anthony C. Woodbury, "What is an Endangered Language?" pamphlet (Washington, DC: Linguistic Society of America, undated); UNESCO Ad Hoc Expert Group on Endangered Languages, "Language Vitality and Endangerment," approved by the participants at International Expert Meeting on the UNESCO Programme Safeguarding of Endangered

Notes

Language, Paris, 10–12 March 2003.

5. K. David Harrison, "Vanishing Voices: What Else is Lost?" *Science & Spirit*, November/December 2004.

6. Andrew Dalby, *Language in Danger* (New York: Columbia University Press, 2003), p. 212.

7. UNESCO, op. cit. note 3; Jessica Ebert, "Tongue Tied," *Nature*, 10 November 2005.

8. Many of these languages had only one or two elderly speakers left when they were catalogued and have probably become extinct in recent years.

9. Gordon, op. cit. note 2.

10. UNESCO Ad Hoc Expert Group, op. cit. note 4.

11. Gordon, op. cit. note 2.

12. Ibid.

13. Ibid.

14. Ibid.

15. Ibid.

16. Terra Lingua Web site, at www.terralingua.org.

17. Gordon, op. cit. note 2.

18. Ibid.

19. Woodbury, op. cit. note 4.

20. Daniel Nettle and Suzanne Romaine, *Vanishing Voices: The Extinction of the World's Languages* (New York: Oxford University Press, 2000), p. 40.

21. UNESCO, op. cit. note 3.

22. Robert E. Moore, "Endangered," *Journal of Linguistic Anthropology*, June-December 1999.

23. "Babel Runs Backwards," op. cit. note 1.

24. Ibid.

25. Dalby, op. cit. note 6, p. 219.

26. Joel Kuipers, professor of Anthropology and International Affairs, George Washington University, Washington DC, discussion with author, 1 February 2006.

27. The Linguasphere Observatory, at www.lingua sphere.org, 2004; estimates of numbers of speakers for individual languages vary greatly.

28. Herman Wasserman, "Between the Local and the Global: South African Languages and the Internet," *African and Asian Studies*, vol. 1, no. 4 (2002), p. 303.

29. British Council Web site, at www.britishcouncil.org/ english/engfaqs.htm; Tan Wei, "Crazy for English," *Beijing Review*, 7 August 2003, p. 18; David Graddol, *English Next* (Plymouth, U.K.: Latimer Trend & Company Ltd. for British Council, 2006), p. 95.

30. Graddol, op. cit. note 29, p. 95.

31. Stephen Phillips, "Ambitious Flock to Chinese Classes," *The Times Educational Supplement*, 28 October 2005, p. 20.

32. UNESCO, op. cit. note 3; Gordon, op. cit. note 2.

33. Wasserman, op. cit. note 28.

34. Mark Abley, "Spoken Here: The Mohawk Community of Kahnawa:ke near Montreal Takes a Page from Quebec's Language Legislation," *Canadian Geographic*, September/October 2003.

35. Ibid.

36. Bill Marsh, "Livonian Spoken Here (for Now)," *New York Times*, 19 June 2005.

37. Ibid.

38. UNESCO, op. cit. note 3.

39. Ebert, op. cit. note 7.

SLUMS GROW AS URBAN POVERTY ESCALATES (pages 114–15)

1. U.N. Department of Economic and Social Affairs, Population Division, *World Urbanization Prospects: The 2003 Revision* (New York: United Nations, 2004), p. 13. As *Vital Signs* went to press, the U.N. Population Division was updating its estimate of the number of people in "urban agglomerations," which typically include a town or city as well as adjacent suburbs, by reviewing census data from the world's nations, which define "urban" in different ways.

2. U.N. Department of Economic and Social Affairs, op. cit. note 1, pp. 13–14, 178–79.

3. UN-HABITAT, *State of the World's Cities 2006/7: The Millennium Development Goals and Urban Sustainability: 30 Years of the Habitat Agenda* (Nairobi: 2006), early draft of final version, reviewed 24 March 2006.

4. Birth of slums in London from Lewis Mumford, *The City in History: Its Origins, Its Transformations, and Its Prospects* (San Diego, CA: Harcourt Brace & Company, 1961), pp. 458–65.

5. Joseph B. Treaster and Kate Zernike, "Hurricane Slams into Gulf Coast; Dozens are Dead," *New York Times*, 30 August 2005; David Gonzalez, "From Margins of Society to Center of the Tragedy," *New York Times*, 2 September 2005.

6. Christopher Cooper, "Old-Line Families Escape the Worst of Flood and Plot the Future," *Wall Street Journal*, 8 September 2005; Michael Lewis, "Wading Toward Home," *New York Times Magazine*, 9 October 2005.

7. "Anarchy-Plagued Mogadishu Empathizes with New Orleans: Residents," *Agence France Presse*, 2 September 2005.

8. "French Riots Enter 10th Night Despite Crackdown Vow," *Agence France Presse*, 5 November 2005.

9. "An Underclass Rebellion—France's Riots," *The Economist*, 12 November 2005.

10. "Immigrants in Paris: Dreams Go Up in Flames," in

UN-HABITAT, op. cit. note 3; "Children Die in Paris Hostel Fire," *BBC News*, 26 August 2005.

11. Figure of 90 percent from UN-HABITAT, op. cit. note 3; television coverage from CNN and BBC.

12. UN-HABITAT, op. cit. note 3; Figure 1 from UN-HABITAT, Global Urban Observatory, 2005.

13. UN-HABITAT, op. cit. note 3.

14. Ibid.

15. Ibid.

16. UN-HABITAT, *State of the World's Cities 2001* (Nairobi: 2001), pp. 116–17.

17. Freedom House, "Combined Average Rankings: Independent Countries 2005," in *Freedom in the World 2005*, at www.freedomhouse.org/template.cfm?page =193&year=2005; U.N. Development Programme, *Human Development Report 2005* (New York: Oxford University Press, 2005), pp. 219–22.

18. UN-HABITAT, op. cit. note 3.

19. Ibid.

20. Ibid.

21. Ibid.

22. Ibid.

23. Ibid.

24. Ibid.

25. Ibid.

26. Ibid.

27. UNAIDS/World Health Organization, *AIDS Epidemic Update*, December 2004, cited in UN-HABITAT, op. cit. note 3.

28. Anna Tibaijuka, United Nations Special Envoy on Human Settlements Issue in Zimbabwe, *Report of the Fact-Finding Mission to Zimbabwe to Assess the Scope and Impact of Operation Murambatsvina* (New York: United Nations, 2005).

29. Ibid.

30. UN-HABITAT, op. cit. note 3.

31. "Beijing Relocating Thousands of Households Ahead of 2008 Olympics," *Associated Press*, 10 March 2004; Centre on Housing Rights and Evictions, "Zimbabwe, China and the State Government of Maharashtra Cited for Severe Human Rights Violations by Housing Rights Group," press release (Geneva: 20 November 2005).

32. United Nations, *Millennium Summit Declaration* (New York: 2000).

33. UN-HABITAT, op. cit. note 3.

34. Ibid.

35. Ibid.

36. Molly O'Meara Sheehan, "Uniting Divided Cities," in Worldwatch Institute, *State of the World 2003* (New York: W. W. Norton & Company, 2003), pp. 130–51.

ACTION NEEDED ON WATER AND SANITATION (pages 116–17)

1. World Health Organization (WHO) and UNICEF Joint Monitoring Program for Water Supply and Sanitation, "Water Supply Data at Global Level," at www.wssinfo.org/en/22_wat_global.html, viewed 16 February; definition from WHO and UNICEF, "The Joint Monitoring Programme: Definitions," at www.wssinfo.org/en/122_definitions.html, viewed 10 February 2006. Updated data on water and sanitation trends were not available when *Vital Signs* went to press; see the Joint Monitoring Programme's Web site for updates.

2. WHO and UNICEF Joint Monitoring Program for Water Supply and Sanitation, "Sanitation Data at Global Level," at www.wssinfo.org/en/32_san_global .html; definition from WHO and UNICEF, "Definitions," op. cit. note 1.

3. WHO and UNICEF, "Water Supply Data," op. cit. note 1; WHO and UNICEF, "Sanitation Data," op. cit. note 2; United Nations, *The Millennium Development Goals Report: 2005* (New York: 2005), p. 34.

4. UN Millennium Project Task Force on Water and Sanitation, *Health, Dignity, and Development: What Will It Take*, abridged ed. (New York: 2005), p. 17.

5. Relative number of deaths from various water-related diseases from Peter H. Gleick, *The World's Water 2004–2005* (Washington, DC: Island Press, 2004), p. 10.

6. WHO, "Facts and Figures: Water, Sanitation and Hygiene Links to Health," at www.who.int/water _sanitation_health/publications/factsfigures04/en.

7. "World Leaders Adopt 'United Nations Millennium Declaration' at Conclusion of Extraordinary Three-Day Summit," press release (New York: United Nations, 8 September 2000).

8. "Plan of Implementation of the World Summit on Sustainable Development," Paragraph 8, at www.un.org/esa/sustdev/documents/WSSD_POI_P D/English/WSSD_PlanImpl.pdf.

9. UNICEF, "Water and Sanitation Coverage," at www.unicef.org/wes/index_statistics.html, viewed 10 February 2006.

10. Ibid.

11. WHO and UNEP, "Water Supply Data," op. cit. note 1.

12. Ibid.

13. United Nations, op. cit. note 3, p. 33.

14. Ibid.

15. Ibid.

16. UNICEF, op. cit. note 9.

17. World Bank, *Global Monitoring Report 2005* (Wash-

ington, DC: 2005), p. 76, based on data from the Joint Monitoring Program for Water Supply and Sanitation of WHO and UNICEF.

18. WHO and UNICEF, "Sanitation Data," op. cit. note 2.

19. Ibid.

20. WHO and UNICEF, *Water for Life: Making it Happen* (Geneva: 2005), p. 36.

21. Ibid., p. 5.

22. Ibid.

23. Ibid.

24. Ibid.

25. UN Millennium Project Task Force on Water and Sanitation, op. cit. note 4, p. 24.

26. Innovations from William K. Reilly and Harriet C. Babbitt, *A Silent Tsunami: The Urgent Need for Clean Water and Sanitation* (Washington, DC: Aspen Institute, 2005), pp. 5–14.

27. UN Millennium Project Task Force on Water and Sanitation, op. cit. note 4, p. 31; WHO and UNICEF, op. cit. note 20, p. 28.

28. WHO and UNICEF, op. cit. note 20, p. 30.

29. Investment needs and paybacks from WHO and UNICEF, op. cit. note 20, p. 4.

30. Ibid.

31. Ibid.

32. Ibid.; for information on the 4th World Water Forum, see www.worldwaterforum4.org.mx/home/home.asp?lan=.

CAR-SHARING CONTINUES TO GAIN MOMENTUM (pages 118–19)

1. Susan Shaheen, Daniel Sperling, and Conrad Wagner, "Carsharing in Europe and North America: Past, Present and Future," *Transportation Quarterly*, vol. 52, no. 3 (1998), p. 35; Susan Shaheen, *Dynamics in Behavioral Adaptation to a Transportation Innovation: A Case Study of CarLink—A Smart Carsharing System* (Davis, CA: Institute of Transportation Studies, 1999), p. 44.

2. Susan Shaheen, Adam Cohen, and J. Darius Roberts, "Carsharing in North America: Market Growth, Current Developments, and Future Potential," *Transportation Research Record: Journal of the Transportation Research Board*, forthcoming.

3. Current estimates of car-sharing members and vehicles in Europe are based on discussions with Michael Glotz-Richter, 27 January 2006, with Conrad Wagner, 12 December 2005, with Daniel Bongardt, 10 January 2006, and with Henry Mentink, 13 January 2006; estimates for North America based on correspondence with 28 North American car-sharing organizations in January 2006; Asian estimates

based on discussions with Ruey Cheu, 12 December 2005, and with Noynoi Fukuda, 15 January 2006; estimates for Australia based on discussions with Nicholas Lowe, 2 December 2005, with Monique Conheady, 13 December 2005, and with Paul Reichman, 14 December 2005; historical estimates from Gary Gardner, "Car-Sharing Emerging," in Worldwatch Institute, *Vital Signs 2002* (New York: W. W. Norton & Company, 2002), pp. 150–51.

4. Lowe, op. cit. note 3; Conheady, op. cit. note 3; Reichman, op. cit. note 3.

5. Sylvia Harms and Bernard Truffer, *The Emergence of a Nationwide Carsharing Co-operative in Switzerland*, prepared for Eidg. Anstalt fur Wasserversorgung und Gewasserschutz, Switzerland, March 1998.

6. Eric Britton, "A Short History of Early Car Sharing Innovations," *Carsharing 2000: Sustainable Transport's Missing Link—Journal of World Transport Policy & Practice*, January 2000, pp. 9–15; Witkar, at home.deds.nl/~quip/deel/witkar.html, viewed 9 January 2006; Steven Cousins, "Theory, Benchmarking, Barriers to Carsharing: An Alternative Vision & History," *Carsharing 2000: Sustainable Transport's Missing Link—Journal of World Transport Policy & Practice*, January 2000, pp. 44–52; Martin Strid, "Sweden–Getting Mobilized," *Carsharing 2000: Sustainable Transport's Missing Link—Journal of World Transport Policy & Practice*, January 2000, pp. 84–90.

7. Shaheen, Sperling, and Wagner, op. cit. note 1, pp. 40–41; Shaheen, op. cit. note 1, pp. 13–16.

8. Shaheen, Sperling, and Wagner, op. cit. note 1, p. 38; Shaheen, op. cit. note 1, p. 48.

9. Christian Ryden and Emma Morin, *Mobility Services for Urban Sustainability: Environmental Assessment*, Report WP 6 (Stockholm, Sweden: Trivector Traffic AB, 2005); "News," Autoshare, at www.autoshare.com/aboutus_news.html, viewed 31 July 2005; Clayton Lane, "Philly CarShare: First-Year Social and Mobility Impacts of Car Sharing in Philadelphia, Pennsylvania," *Transportation Research Record: Journal of the Transportation Research Board, No. 1927*, 2005, pp. 158–66; "Zipcar Customer Survey Shows Car-Sharing Leads to Car Shedding," at www.zipcar.com/press/releases/press-21, viewed 31 July 2005; "Impact," Flexcar at www.flexcar.com/vision/impact.asp, viewed 31 July 2005.

10. Robert Benoit, "Potentiel de L'Auto-Partage Dans Le Cadre d'Une Politique de Gestion de La Demande en Transport," *Forum de L'AQTR, Gaz à Effet de Serre: Transport et Développement, Kyoto: Une Opportunité d'Affaires?* (Montreal, PQ, Canada: Communauto, Inc., 2000); Nicole Jensen, *The Co-operative Auto*

Network Social and Environmental Report 2000–2001 (Vancouver, BC, Canada: 2001); "News," op. cit. note 9; Lane, op. cit. note 9; Jeff Price and Chris Hamiliton, *Arlington Pilot Carshare Program: First-Year Report* (Arlington, VA: Arlington County Commuter Services, Division of Transportation, Department of Environmental Services, 2005); Richard Katzev, *Carsharing Portland: Review and Analysis of Its First Year* (Portland, OR: Department of Environmental Quality, 1999).

11. Conrad Wagner, ATG-UMFRAGE 1990, ATG, Stans, Germany, 1990, cited in Peter Muheim and Partner, *Car Sharing Studies: An Investigation*, prepared for the Graham Lightfoot Study (Scariff, Ireland: Taylor Lightfoot, Transport Consultants, 1996); Herbert Baum and Stephan Pesch, "Untersuchung der Eignung von Carsharing im Hinblick auf die Reduzierung von Stadtverkehrsproblemen," *Bundesministerium fur Verkehr* (Bonn, Germany: 1994).

12. U.S. Department of Transportation, Bureau of Transportation Statistics, *National Household Transportation Survey 2001 Highlights Report* (Washington, DC: Bureau of Transportation Statistics, 2003).

13. Ryden and Morin, op. cit. note 9, p. 19.

14. Lane, op. cit. note 9; "Zipcar Customer Survey," op. cit. note 9; Georgia Cooper, Deborah Howe, and Peter Mye, *The Missing Link: An Evaluation of Car-Sharing Portland Inc.* (Portland, OR: Department of Environmental Quality, 2000); "First-Ever Study of Car-Sharing," news release (San Francisco: City Car-Share, 7 January 2004).

15. Lane, op. cit. note 9; "Zipcar Customer Survey," op. cit. note 9; Cooper, Howe, and Mye, op. cit. note 14; "First-Ever Study of Car-Sharing," op. cit. note 14.

16. Author's estimate based on Lane, op. cit. note 9, on "Zipcar Customer Survey," op. cit. note 9, on Cooper, Howe, and Mye, op. cit. note 14, and on "First-Ever Study of Car-Sharing," op. cit. note 14.

17. Ryden and Morin, op. cit. note 9, p. 22.

18. Jensen, op. cit. note 10; "First-Ever Study of Car-Sharing," op. cit. note 14; Elizabeth Reynolds and Kevin McLaughlin, "The Smart Alternative to Owning a Car," brochure (Toronto, ON, Canada: AutoShare, 2001).

19. Susan Shaheen, Mollyanne Meyn, and Kamill Wipyewski, "U.S. Shared-Use Vehicle Survey Findings on Carsharing and Station Car Growth: Obstacles and Opportunities," *Transportation Research Record: Journal of the Transportation Research Board, No. 1841*, 2003, pp. 90–98; Todd Litman, "Evaluating Carsharing Benefits," *Transportation Research Record, Journal of the Transportation Research Board, No. 1702*, 2000, pp. 31–35.

20. Reynolds and McLaughlin, op. cit. note 18; Litman, op. cit. note 19, p. 31; "Carsharing," Calgary Alternative Transportation Cooperative, at www.catco-op.org/carsharing.html.

21. Susan Shaheen, Andrew Schwartz, and Kamill Wipyewski, "Policy Considerations for Carsharing and Station Cars: Monitoring Growth, Trends, and Overall Impacts," *Transportation Research Record: Journal of the Transportation Research Board, No. 1887*, 2004, pp. 128–36.

22. Current developments are based on communication with Michael Glotz-Richter, 9 December 2005, and on various experts cited in note 3.

23. Future developments are based on communication with Glotz-Richter, op. cit. note 22, and on various experts cited in note 3.

OBESITY REACHES EPIDEMIC LEVELS
(pages 120–21)

1. World Health Organization (WHO), *Obesity and Overweight* (Geneva: 2003).

2. U.S. Centers for Disease Control and Prevention (CDC), *Trends Data: Obesity by BMI* (Atlanta, GA: November 2003); 40 percent from WHO, *Global Infobase Online*, at infobase.who.int.

3. "Chinese Concern at Obesity Surge," *BBC News*, 12 October 2004; the Chinese government now defines obesity according to a non-standard body mass index threshold.

4. WHO, op. cit. note 2.

5. U.N. Food and Agriculture Organization (FAO), "The Nutrition Transition and Obesity," in *The Developing World's New Burden: Obesity* (Rome: 2002); Ania Lichtarowicz, "Obesity Epidemic 'Out of Control'," *BBC News*, 31 October 2004.

6. Meat production and sugar consumption from FAO, *FAOSTAT Statistical Database*, at apps.fao.org.

7. In 1962, a diet containing 20 percent of total energy from fat correlated with a per capita gross national product (GNP) of $1,475. By 1990, a per capita GNP of just $750 correlated with the same diet. FAO, op. cit. note 5.

8. Gary Gardner, "People Everywhere Eating More Fast Food," in Worldwatch Institute, *Vital Signs 1999* (New York: W. W. Norton & Company, 1999), pp. 150–51; Erik Assadourian, "Soda," in Worldwatch Institute, *State of the World 2004* (New York: W. W. Norton & Company, 2004), p. 94.

9. S. H. Holt et al., "A Satiety Index of Common Foods," *European Journal of Clinical Nutrition*, September 1995, pp. 675–90; FAO, op. cit. note 5.

10. Lester Brown, "Obesity Epidemic Threatens Health

in Exercise-Deprived Societies," *Earth Policy Alerts* (Washington, DC: Earth Policy Institute, 19 December 2000).

11. Peter G. Kopelman, "Obesity as a Medical Problem," *Nature*, 6 April 2000, p. 636; WHO, "Children, Mobility and Environmental Health," in *Children's Environmental Health* (Geneva: 2006); Robert Kubey and Mihaly Csikszentmihalyi, "Television Addiction Is No Mere Metaphor," *Scientific American,* February 2002, pp. 74–80.

12. Barry M. Popkin, "Urbanization and the Nutrition Transition," in *Achieving Urban Food and Nutrition Security in the Developing World, A 2020 Vision for Food, Agriculture, and the Environment*, Focus 3, Brief 7 (Washington, DC: International Food Policy Research Institute, 2000).

13. Adam Drewnowski and S. E. Specter, "Poverty and Obesity: The Role of Energy Density and Energy Costs," *American Journal of Clinical Nutrition*, January 2004, pp. 6–16.

14. New York City Department of Health and Mental Hygiene, "3.2 Million New Yorkers Are Overweight or Obese," press release (New York: 22 November 2005); DW-World.De, "East Germans Tipping the Scales" (Deutsche Welle: 2005).

15. Society for Women's Health Research, "Sex Differences in Obesity," press release (Washington, DC: 4 May 2005); Rolando Fuertes, Jr., "Obesity in Arab World Is on Rise," *The Seoul Times*, 7 February 2006.

16. James Bindon and Paul Baker, "Bergmann's Rule and the Thrifty Genotype," *American Journal of Physical Anthropology*, October 1997, pp. 201–10.

17. Ibid.; Figure 1 from WHO, op. cit. note 2.

18. WHO Expert Consultation, "Appropriate Body-mass Index for Asian Populations and Its Implications for Policy and Intervention Strategies," *Lancet*, 10 January 2004, pp. 157–63.

19. Ian Janssen, Peter T. Katzmarzyk, and Robert Ross, "Waist Circumference and Not Body Mass Index Explains Obesity-related Health Risk," *American Journal of Clinical Nutrition*, March 2004, pp. 379–84.

20. WHO, op. cit. note 1; respiratory diseases from CDC, "Overweight and Obesity: Frequently Asked Questions," fact sheet (Atlanta, GA: 18 November 2005).

21. WHO, op. cit. note 1; CDC, op. cit. note 21; absenteeism from American Obesity Association, "Costs of Obesity," 2 May 2005, at www.obesity.org.

22. International Diabetes Federation, "About Diabetes," 2003, at www.idf.org; Laura Wood, "Prevalence of Diabetes Set to Increase from 177m Worldwide to 370m by 2030," *Medical News Today,* 9 April 2005.

23. American Obesity Association, op. cit. note 21.

24. Figure of 33 billion euros from James Fry and Willa Finley, "The Prevalence and Cost of Obesity in the E.U.," *Proceedings of the Nutrition Society*, August 2005, pp. 359–362(4); up to 130 billion euros from Geof Rayner and Mike Rayner, "Fat Is An Economic Issue!" *Eurohealth*, spring 2003.

25. Figure of $75 billion from Eric Finkelstein, Ian Fiebelkorn, and Guijing Wang, "State-Level Estimates of Annual Medical Expenditures Attributable to Obesity," *Obesity Research*, January 2004, pp. 18–24; $118 billion from CDC, *Preventing Obesity and Chronic Diseases Through Good Nutrition and Physical Activity*" (Atlanta: GA: July 2005).

26. U.S. National Institutes of Health, "Obesity Threatens to Cut U.S. Life Expectancy," press release (Bethesda, MD: 16 March 2005).

27. American Obesity Association, op. cit. note 21.

28. Laura Cummings, "The Diet Business: Banking on Failure," *BBC News Online*, 5 February 2003; J. M. Friedman, "Obesity in the New Millennium," *Nature*, 6 April 2000, pp. 632–34.

29. Steve Bloomfield, "Ops for Obesity Double As Epidemic Grows," *Independent* (London), 4 September 2005.

30. British Medical Association, *Preventing Childhood Obesity* (London: June 2005); American Obesity Association, "Childhood Obesity," 2 May 2005, at www.obesity.org.

31. Cheong Mui Toh, Jeffery Cutter, and Suok Kai Chew, "School Based Intervention Has Reduced Obesity in Singapore," *British Medical Journal*, 16 February 2002, p. 427.

32. The National Academies, "Food Marketing Aimed at Kids Influences Poor Nutritional Choices, IOM Study Finds," press release (Washington, DC: 6 December 2005).

33. Peter Ford, "Foes of 'Globesity' Run Afoul of Sugar's Friends," *Christian Science Monitor*, 19 February 2004.

34. "Governors Urge Change in Eating Culture," *Associated Press*, 25 February 2006; William Mortiz, "A Survey of North American Bicycle Commuters," 24 January 1997, at www.bicyclelife.com.

35. School of Public Health, University of North Carolina, "Primer on Active Living By Design," at www.activelivingbydesign.org, viewed 6 February 2006.

CORPORATE RESPONSIBILITY REPORTS TAKE ROOT (pages 122–23)

1. Number reporting from Paul Scott, CorporateRegister.com, e-mail to author, 23 January 2006.

2. Ibid.

3. Ibid.
4. Paul Scott, CorporateRegister.com, e-mail to author, 31 August 2005.
5. University of Amsterdam and KPMG Global Sustainability Services, *KPMG International Survey of Corporate Responsibility Reporting 2005* (Amsterdam: 2005).
6. Ibid.
7. "Non-financial Reporting Status of the FTSE 100," at www.corporateregister.com/charts/FTSE.htm, viewed 3 January 2006.
8. Ibid.
9. "Socially Responsible Investment Analysts Find More Large U.S. Companies Reporting on Social and Environmental Issues," *Environmental News Network*, 13 July 2005.
10. "Survey: Sustainability Gaining Prominence in Annual Reports," *GreenBiz.com*, 6 October 2005.
11. Scott, op. cit. note 1; number of transnational corporations from U.N. Conference on Trade and Development, *World Investment Report 2005* (New York: 2005), pp. 264–65.
12. ACCA and CorporateRegister.com, *Towards Transparency: Progress on Global Sustainability Reporting 2004* (London: Certified Accountants Educational Trust, 2004); Standard & Poor's, SustainAbility, and U.N. Environment Programme (UNEP), *Risk & Opportunity, The Global Reporters 2004 Survey of Corporate Sustainability Reporting: Best Practice in Non-Financial Reporting* (London: 2004).
13. Standard & Poor's, SustainAbility, and UNEP, op. cit. note 12, p. 5.
14. ACCA and CorporateRegister.com, op. cit. note 12, p. 8.
15. Scott, op. cit. note 4.
16. Standard & Poor's, SustainAbility, and UNEP, op. cit. note 12, p. 15.
17. The Climate Group, *Carbon Down, Profit Up* (London: 2005).
18. Ibid.
19. *Striking a Balance: Corporate Social Responsibility Fiscal 2004 Annual Report* (Seattle: Starbucks Coffee Company, 2005) (note that 2004 was the first year that Starbucks started reporting and verifying total coffee purchasing).
20. Ibid.
21. Goal for 2007 from Sue Mecklenburg, vice president, Corporate Social Responsibility, Starbucks Coffee Company, e-mail to author, 28 September 2005.
22. Toyota Motor Company, *Environmental and Social Report 2005* (Toyota City, Japan: 2005), p. 82.
23. Ibid., pp. 14–15.

24. Ibid.
25. Global Reporting Initiative, "Reporter Statistics," database at www.globalreporting.org/guidelines/ReportersStats.xls, updated 4 January 2006; 300 from Scott, op. cit. note 1.
26. "GRI at a Glance," at www.globalreporting.org/about/brief.asp, viewed 3 January 2006; "Lighthouse G3: Third Generation GRI Guidelines Shine a Beacon for Sustainability Reporters," *SocialFunds.com*, 30 September 2005.
27. "G3: Major Goals," at www.grig3.org/mayorgoals.html, viewed 3 January 2006.
28. Ibid.
29. Scott, op. cit. note 1.
30. Lucien J. Dhooge, "Beyond Voluntarism: Social Disclosure and France's Nouvelles Régulations Économiques," *Arizona Journal of International & Comparative Law*, vol. 21, no. 2 (2004), pp. 441–90.
31. Ibid.
32. Ibid.
33. Ibid.
34. U.K. Department of Trade and Industry, Operating and Financial Review, at www.dti.gov.uk/cld/financialreview.htm, viewed 15 January 2006.
35. Standard & Poor's, SustainAbility, and UNEP, op. cit. note 12, p. 7.
36. William Baue, "UK Kills Operating and Financial Review of Environmental and Social Information," SocialFunds.com, 8 December 2005.

NANOTECHNOLOGY TAKES OFF (pages 124–25)

1. David Rejeski, Director, Project on Emerging Nanotechnologies, Woodrow Wilson International Center for Scholars, Testimony before U.S. House of Representatives, Hearing on Environmental and Safety Impacts of Nanotechnology, 17 November 2005.
2. Stacy Lawrence, "Nanotech Grows Up," *Technology Review*, June 2005.
3. Karen Breslau (with Joanna Chung), "Big Future in Tiny Spaces," *Newsweek*, 23 September 2002.
4. *The National Nanotechnology Initiative: Research and Development Leading to a Revolution in Technology and Industry, Supplement to the President's FY 2006 Budget* (Washington, DC: 2005), in current dollars; for nanotech funding to date, see President's Council of Advisors on Science and Technology, *The National Nanotechnology Initiative at Five Years: Assessment and Recommendations of the National Nanotechnology Advisory Panel* (Washington, DC: 2005).
5. See, for example, M. Roco and W. Bainbridge, eds., *Converging Technologies for Improving Human Perfor-*

mance (Netherlands: Kluwer Academic Publishers, 2003).

6. Chunli Bai, "Ascent of Nanoscience in China," *Science*, 1 July 2005, p. 62.

7. Steve Jurvetson, "Transcending Moore's Law with Molecular Electronics," *Nanotechnology Law & Business Journal*, vol. 1, no. 1, article 9, p. 9.

8. Mihail Roco, Senior Adviser for Nanotechnology, at the National Science Foundation, has estimated the number of nano-products to be 700 (discussion with authors, 30 September 2005). More recently, Cate Alexander, Communications Director at the National Nanotechnology Coordination Office, estimated that the number of nanotech-related products—including raw materials, electronics components and research, process, and software tools—is 600 (e-mail to authors, 18 January 2006). Determining the number of products is a difficult task, as there are no labeling requirements or other regulations for nanoscale materials.

9. See, for example, Royal Society and Royal Academy of Engineering, *Nanoscience and Nanotechnologies: Opportunities and Uncertainties* (London: 2004).

10. Ibid., Chapter 5.

11. For an introduction to synthetic biology and its possible applications, see Andrew Pollack, "Custom-Made Microbes, at Your Service," *New York Times*, 17 January 2006.

12. See International Center for Bioethics, Culture and Disability, at www.bioethicsanddisability.org.

13. The South Centre, *The Potential Impacts of Nano-Scale Commodities on Commodity Markets: The Implications for Commodity Dependent Developing Countries*, prepared by ETC Group, Trade-Related Agenda, Development and Equity, Research Paper No. 4 (Geneva: 2005).

14. ETC Group, *Nanotech's "Second Nature" Patents: Implications for the Global South* (Ottawa, ON, Canada: 2005).

The Vital Signs Series

Some topics are included each year in *Vital Signs*; others are covered only in certain years. The following is a list of topics covered in *Vital Signs* thus far, with the year or years they appeared indicated in parentheses. Those marked with a bullet (•) appeared in Part One, which includes time series of data on each topic; 2006 indicates this edition of *Vital Signs*.

AGRICULTURE AND FOOD

Agricultural Resources
- •Fertilizer Use (1992–2001)
- •Grain Area (1992–93, 1996–97, 1999–2000)
- •Grain Yield (1994–95, 1998)
- •Irrigation (1992, 1994, 1996–99, 2002)
  Livestock (2001)
  Organic Agriculture (1996, 2000)
  Pesticide Control or Trade (1996, •2000, 2002, •2006)
  Transgenic Crops (1999–2000)
  Urban Agriculture (1997)

Food Trends
- •Aquaculture (1994, 1996, 1998, 2002, 2005)
  Biotech Crops (2001–02)
- •Cocoa Production (2002)
- •Coffee (2001)
- •Fish (1992–2000, 2006)
- •Grain Production (1992–2003, 2005–06)
- •Grain Stocks (1992–99)
- •Grain Used for Feed (1993, 1995–96)
- •Meat (1992–2000, 2003, 2005–06)
- •Milk (2001)
- •Soybeans (1992–2001)
- •Sugar and Sweetener Use (2002)

THE ECONOMY

Resource Economics
  Agricultural Subsidies (2003)
- •Aluminum (2001, 2006)
  Arms and Grain Trade (1992)
  Commodity Prices (2001)
  Fossil Fuel Subsidies (1998)
- •Gold (1994, 2000)
  Illegal Drugs (2003)
  Metals Exploration (1998, •2002)
- •Metals Production (2002)
- •Paper (1993, 1994, 1998–2000)
  Paper Recycling (1994, 1998, 2000)
- •Roundwood (1994, 1997, 1999, 2002, 2006)
  Seafood Prices (1993)
- •Steel (1993, 1996, 2005–06)
  Steel Recycling (1992, 1995)
  Subsidies for Environmental Harm (1997)
  Wheat/Oil Exchange Rate (1992–93, 2001)

World Economy and Finance

- Agricultural Trade (2001)
  Aid for Sustainable Development (1997, 2002)
- Developing-Country Debt (1992–95, 1999–2003)
  Environmental Taxes (1996, 1998, 2000)
  Food Aid (1997)
- Global Economy (1992–2003, 2005–06)
  Microcredit (2001)
- Oil Spills (2002)
  Private Finance in Third World (1996, 1998, 2005)
  R&D Expenditures (1997)
  Socially Responsible Investing (2001, 2005)
  Stock Markets (2001)
- Trade (1993–96, 1998–2000, 2002, 2005)
  Transnational Corporations (1999–2000)
- U.N. Finances (1998–99, 2001)

Other Economic Topics

- Advertising (1993, 1999, 2003, 2006)
  Charitable Donations (2002)
  Cigarette Taxes (1993, 1995, 1998)
  Corporate Responsibility (2006)
  Cruise Industry (2002)
  Ecolabeling (2002)
  Government Corruption (1999, 2003)
  Health Care Spending (2001)
  Nanotechnology (2006)
  Pay Levels (2003)
  Pharmaceutical Industry (2001)
  PVC Plastic (2001)
  Satellite Monitoring (2000)
- Television (1995)
  Tourism (•2000, •2003, 2005)

ENERGY AND ATMOSPHERE

Atmosphere

- Carbon Emissions (1992, 1994–2002)
- Carbon and Temperature Combined (2003, 2005–06)
- CFC Production (1992–96, 1998, 2002)
- Global Temperature (1992–2002)
  Weather-related Disasters (•1996–2000, 2001, 2003, •2005–06)

Fossil Fuels

- Carbon Use (1993)
- Coal (1993–96, 1998)
- Fossil Fuels Combined (1997, 1999–2003, 2005–06)
- Natural Gas (1992, 1994–96, 1998)
- Oil (1992–96, 1998)

Renewables, Efficiency, Other Sources

- Biofuels (2005–06)
- Compact Fluorescent Lamps (1993–96, 1998–2000, 2002)
- Efficiency (1992, 2002, 2006)
- Geothermal Power (1993, 1997)
- Hydroelectric Power (1993, 1998, 2006)
- Nuclear Power (1992–2003, 2005–06)
- Solar Cells (1992–2002, 2005–06)
- Wind Power (1992–2003, 2005–06)

THE ENVIRONMENT

Animals

Amphibians (1995, 2000)
Aquatic Species (1996, 2002)
Birds (1992, 1994, 2001, 2003, 2006)
Mammals (2005)
Marine Mammals (1993)
Primates (1997)
Vertebrates (1998)

Natural Resource Status

Coral Reefs (1994, 2001, 2006)
Farmland Quality (2002)
Forests (1992, 1994–98, 2002, 2005–06)
Groundwater (2000, 2006)
Ice Melting (2000, 2005)
Mangroves (2006)
Ozone Layer (1997)
Plant Diversity (2006)
Water Scarcity (1993, 2001–02)

Water Tables (1995, 2000)
Wetlands (2001, 2005)

Natural Resource Uses

Biomass Energy (1999)
Dams (1995)
Ecosystem Conversion (1997)
Energy Productivity (1994)
Organic Waste Reuse (1998)
Soil Erosion (1992, 1995)
Tree Plantations (1998)

Pollution

Acid Rain (1998)
Air Pollution (1993, 1999, 2005)
Algal Blooms (1999)
Hazardous Wastes (2002)
Lead in Gasoline (1995)
Mercury (2006)
Nuclear Waste (1992, ●1995)
Pesticide Resistance (●1994, 1999)
● Sulfur and Nitrogen Emissions (1994–97)

Other Environmental Topics

Environmental Indicators (2006)
Environmental Treaties (●1995, 1996, 2000, 2002)
Nitrogen Fixation (1998)
Pollution Control Markets (1998)
Sea Level Rise (2003)
Semiconductor Impacts (2002)
Transboundary Parks (2002)
● World Heritage Sites (2003)

THE MILITARY

● Armed Forces (1997)
Arms Production (1997)
● Arms Trade (1994)
Landmines (1996, 2002)
● Military Expenditures (1992, 1998, 2003, 2005–06)
● Nuclear Arsenal (1992–96, 1999, 2001, 2005)

Peacekeeping Expenditures (1993, ●1994–2003, ●2005–06)
Resource Wars (2003)
● Wars (1995, 1998–2003, 2005–06)
Small Arms (1998–99)

SOCIETY and HUMAN WELL-BEING

Health

● AIDS/HIV Incidence (1994–2003, 2005–06)
Alternative Medicine (2003)
Asthma (2002)
Breast and Prostate Cancer (1995)
● Child Mortality (1993, 2006)
● Cigarettes (1992–2001, 2003, 2005)
Drug Resistance (2001)
Endocrine Disrupters (2000)
Food Safety (2002)
Hunger (1995)
● Immunizations (1994)
● Infant Mortality (1992)
Infectious Diseases (1996)
Life Expectancy (1994, ●1999)
Malaria (2001)
Malnutrition (1999)
Mental Health (2002)
Mortality Causes (2003)
Noncommunicable Diseases (1997)
Obesity (2001, 2006)
● Polio (1999)
Sanitation (1998)
Soda Consumption (2002)
Traffic Accidents (1994)
Tuberculosis (2000)
Water and Sanitation (1995, 2006)

Reproduction and Women's Status

Family Planning Access (1992)
Female Education (1998)
Fertility Rates (1993)
Maternal Mortality (1992, 1997, 2003)
● Population Growth (1992–2003, 2005–06)

Sperm Count (1999)
Violence Against Women (1996, 2002)
Women in Politics (1995, 2000)

Social Inequities

Homelessness (1995)
Income Distribution (1992, 1995, 1997, 2002–03)
Language Extinction (1997, 2001, 2006)
Literacy (1993, 2001)
Prison Populations (2000)
Slums (2006)
Social Security (2001)
Teacher Supply (2002)
Unemployment (1999, 2005)

Other Social Topics

Aging Populations (1997)
Fast-Food Use (1999)
International Criminal Court (2003)
Millennium Development Goals (2005)
Nongovernmental Organizations (1999)
Orphans Due to AIDS Deaths (2003)
Public Policy Networks (2005)
Quality of Life (2006)
Refugees (•1993–2000, 2001, 2003, •2005)
Religious Environmentalism (2001)
Urbanization (•1995–96, •1998, •2000, 2002)
Voter Turnouts (1996, 2002)
Wind Energy Jobs (2000)

TRANSPORTATION AND COMMUNICATIONS

- Air Travel (1993, 1999, 2005–06)
- Automobiles (1992–2003, 2005–06)
- Bicycles (1992–2003, 2005–06)
  Car-sharing (2002, 2006)
  Computer Production and Use (1995)
  Gas Prices (2001)
  Electric Cars (1997)
- Internet (1998–2000, 2002)

- Internet and Telephones Combined (2003, 2006)
- Motorbikes (1998)
- Railroads (2002)
- Satellites (1998–99)
- Telephones (1998–2000, 2002)
  Urban Transportation (1999, 2001)